WORKING IN INTERNATIONAL DEVELOPMENT AND HUMANITARIAN ASSISTANCE

An indispensable career guide for everyone wanting to work in or already working in the sector, *Working in International Development and Humanitarian Assistance:*

- provides a general introduction and insight into the sector, for those exploring it as a potential career;
- gives graduates or career changers who are new to the sector an understanding of the skills and experience that will make them stand out from the competition;
- provides practical advice on volunteering and internships, choosing Master's courses and the job search – and what to do when you don't feel you are getting anywhere;
- enables those already working in the sector to gain a long-term view of where they want to go and how they might structure their professional development to gain the skills and competencies necessary to move up the ladder;
- includes case studies, experience and advice from over 100 people already working in the sector, including 50 different areas of speciality.

With case studies, worksheets, CV advice, job profiles and testimonials from people in the field, *Working in International Development and Humanitarian Assistance* provides a refreshingly honest introduction to the field for those exploring it as a potential career. How do you become a country director for an international NGO? How can you become a gender mainstreaming expert? What can you do to get into consultancy?

The book also offers a detailed account of the myriad careers and specialisms available within the sector and methodologically describes the pros and cons of each option. If you are not sure where you want to go with your career, this is the book for you. Whatever your dream job, be it Programme Management, Environmental Advisor, on the ground or in HQ, this book will give you an insight into what working in the sector entails and how you can get into it. It will be an invaluable guide to all readers, irrespective of their country of origin, who are interested in the sector.

Maïa Gedde studied Biological Sciences at Oxford University, but upon graduating, decided there was more to life than cells. A year of travelling and a few temporary jobs later she landed in the Africa Great Lakes and Horn Department of the UK Department for International Development (DFID), which placed development and humanitarian work on her radar. After a short course in Development Studies at Birkbeck College, UK, she went on to do a Master's in Development Studies in Uppsala, Sweden. She co-authored *Working in International Health* (2011), a guide for health professionals. In her early career she coordinated projects in a number of countries including Ghana, Uganda, Malawi and Morocco. She is currently country manager for SPARK, a Dutch NGO working in entrepreneurship and job creation in post-conflict countries.

'A tremendous resource for all those seeking to enter and build their careers in the international development and humanitarian assistance fields. Gedde's book provides much needed advice on the range and diversity of roles, how to identify which aspect of the sector is for you, build your knowledge, skills and evidence, and network and market yourself effectively to find and secure an opportunity. Whether new to the sector, transitioning into it mid-career, or thinking about how to take your experience further, the multitude of examples and case studies woven throughout Gedde's chapters provide superb insights and context to her clear, practical and thorough guidance.'

Dr Jane Chanaa, Careers Team Leader, University of Oxford, UK

'A thoroughly practical guide for anyone considering a career within international development and humanitarian work; it's sure to help you find the right path to make your heart sing.'

Susan Davis, President and CEO, BRAC USA

'This excellent guide comes out at exactly the right time. It provides historical and political context for current discussions around a new set of Sustainable Development Goals, to take us from 2015 to 2030, and provides some wonderfully practical guidance and advice to those who think they might want to work in this area.

The world of international development and humanitarian assistance is complex. It can also be frustrating for practitioners, as some of the stories in this book make clear. Idealism has to be tempered by reality – but it is the best place to start. So if eliminating absolute poverty, protecting people in humanitarian situations and saving the Planet – or any one of the above – sounds as if it might be for you, and you want to know how to get involved, then read on.'

Professor Myles Wickstead CBE, Former Head of Secretariat
to the Commission for Africa (CfA)

WORKING IN INTERNATIONAL DEVELOPMENT AND HUMANITARIAN ASSISTANCE

A career guide

Maïa Gedde

Routledge
Taylor & Francis Group

LONDON AND NEW YORK

First published 2015
by Routledge
2 Park Square, Milton Park, Abingdon, Oxon OX14 4RN

and by Routledge
711 Third Avenue, New York, NY 10017

Routledge is an imprint of the Taylor & Francis Group, an informa business

© 2015 Maïa Gedde

British Library Cataloguing in Publication Data
A catalogue record for this book is available from the British Library

Library of Congress Cataloging in Publication Data
Gedde, Maïa.
  Working in international development and humanitarian assistance:
  a career guide/Maïa Gedde.
  pages cm
  Includes bibliographical references and index.
  1. Humanitarian assistance. 2. Economic assistance. 3. International
  agencies – Officials and employees. 4. Nonprofit organizations –
  Employees. 5. Non-governmental organizations – Employees.
  6. Vocational guidance. I. Title.
  HV553.G43 2015
  361.2'5023 – dc23
  2014036976

ISBN: 978-0-415-69834-4 (hbk)
ISBN: 978-0-415-69835-1 (pbk)
ISBN: 978-0-203-50270-9 (ebk)

Typeset in Bembo and Stone Sans
by Florence Production Ltd, Stoodleigh, Devon, UK

MIX
Paper from
responsible sources
FSC
www.fsc.org   FSC® C013056

Printed and bound in Great Britain by
TJ International Ltd, Padstow, Cornwall

For Taïa, for making me laugh every day
and tolerating my absences.

And my parents, for global and
unconventional foundations.

CONTENTS

ILLUSTRATIONS

Figures

Plates

Tables

CAREER PROFILES

FOREWORD

Reading Maïa Gedde's wonderful guide to working in international development brought home to me how lucky I am. And how old. I started working in development NGOs in the late 1990s, after a 20-year random walk involving a physics degree, backpacking, human rights activism, journalism, a Latin American thinktank and a spell keeping London nuclear free. Things are a bit better organised now – I probably wouldn't give the younger me a job.

The aid business has professionalised over the last few decades: aid agencies have grown enormously in size and sophistication, with a rise in specialisation (humanitarian emergencies, advocacy and campaigns, long-term development, new academic disciplines). It has internationalised, with the old domination by 'white men in shorts' giving way to a much more global intake of personnel. And it has prompted a boom in students seeking qualifications and ways to find that cherished job where you can get paid (a bit) for changing the world.

But that is where international development has so far failed. It has not put in place the kind of entry schemes (graduate entry, sponsored degrees, professional qualifications etc.) and career ladders that other, more established professions have introduced.

In part, I am glad – there is something about making development too slick, too much of a formalised career that could undermine the political and moral basis for getting involved in the first place. Excessive professionalisation could exacerbate the current tendency to try and distil development into an apolitical technocratic exercise, when the reality, whether a country or community prospers or languishes, is determined above all by issues of power and political struggle.

But even if a conveyor belt from university to country director might be a bad idea, it is still worth helping those desperate to get a foot on the ladder before they become disillusioned and drift off to other destinies, and this book makes a real contribution. Not only does it chart the full range of potential jobs and their concomitant lifestyles, it also illustrates it with hundreds of quotes from the men and women who are currently doing them – the whole aid business becomes humanised along the way. Gedde helps the reader sift through the options, finding those that most fit their character and expectations. She even throws in a handy dummy's guide to theory and practice in development.

For years, I have felt a slight twinge of guilt at the inadequacy of the advice I have dispensed to bright-eyed graduates asking how they can get a start in the aid business. Now I know exactly what to recommend, and for that, I am very grateful.

Duncan Green
Senior Strategic Advisor, Oxfam and author of
From Poverty to Power (book and blog)

WHO IS THIS BOOK FOR?

In the last two decades the number of people interested in pursuing a career in the fields of development and humanitarian assistance has grown exponentially. One of the development veterans I spoke to said 'When I entered the field no one wanted to go to work in a place like West Africa. I was one of just a handful of candidates'. These days there are often over 200 applicants per post. Equally, Development Studies Master's courses are now commonplace in universities around the world, whereas two decades ago only a few of select universities offered such courses.

As the quest to end poverty remains at the centre of the media limelight, young people from developing and developed countries alike flock to the challenge. Regardless of academic background they turn towards the sectors of development and humanitarian assistance in search of a dynamic and multidimensional career – often more attractive and exciting than some of the other more conventional employment options. The question for them is how to launch a career with relatively little experience and become a professional in these increasingly competitive sectors.

Mid-career professionals from other fields, in search of a more meaningful and socially responsible career or keen to pursue personal growth and new professional horizons, are also interested in exploring this transition. As the sectors are increasingly valuing corporate experience, this transition is becoming easier. They want to explore what skills are in demand, and how they can market these in a way that is attractive to recruiters in the sector.

Those towards the end of their career also find exciting opportunities to put their skills and knowledge to good use, in new environments that greatly value their extensive skills. What options are there?

So whether you are a newcomer, exploring the field for the first time, a graduate with experience, career changer, sector changer, social entrepreneur or retiree looking to put skills to good use, this book will address many of the questions that you may have.

STRUCTURE OF THE BOOK

This book is grouped into four parts. Parts 1–3 comprise four chapters each. The fourth part focuses on 54 different areas of speciality featuring a personal case study and career trajectory within each. While it is important to understand the whole context, it is not necessary to read the book from beginning to end in order. Each chapter has been designed to be stand-alone, for you to dip in and out of as relevant.

For the newcomer who has a vague notion that they would like to work in either the development or humanitarian sectors, Part 1 helps to shed some light on what work in the sectors really entails and provides a dose of reality. It will also help you to develop more focus and explore what appeals the most.

The broad scope and lack of clear career paths within the sectors can create a confusing and seemingly impenetrable mesh for the job seeker, disillusioned by job applications sent out without receiving any positive response. Part 2 of this book exists to help job seekers and career changers to make concrete plans on how to enter the field, building up relevant skills en route.

For those already working in the sector and planning the next steps in their career, Part 3 looks at some of the different options available. This section helps you to draw on practical advice from senior peers on how to get where you want to go.

Part 4 looks at the different areas of speciality within the sector. This will allow the student to plan their entry better, according to in-demand skills. The career changer will be able to understand what transferable skills they have and how these might be able to be put to good use.

NOTE ON TERMINOLOGY

There is significant controversy around using the term *developing country* – as we may argue, all countries are developing and changing. No country is free from poverty and hardship. However, it is also a term that is widely used, although definitions vary and each organisation tends to have its own. For example, when the Organisation for Economic Co-operation and Development (OECD) uses the term without qualification it is generally taken to mean a country that is eligible for Official Development Assistance (ODA) (see Table 4.2). The World Bank usually uses the term to refer to low- and middle-income countries, assessed by reference to per capita gross national income (GNI). Fortunately, however, the differences in coverage are usually minor.

For the purpose of clarity, the term *international development* has been used to refer to work that is done either a) outside of one's own country of origin or b) in one's country of origin but working for an international organisation. What would be termed local development (working on a local project, in one's own country of origin with local funding streams) is not explicitly included.

ACKNOWLEDGEMENTS

I am extremely grateful to all the friends, colleagues, acquaintances and all those who enthusiastically responded to my emails after a request for information. Almost one hundred people from every corner of the globe have contributed to this book, adding to it by openly sharing their experiences, providing information, reviewing chapters or connecting me to other relevant people. It is truly a collaborative effort and their support and encouragement has helped to make this guide more real; although I alone am responsible for any shortcomings. Some have preferred to remain anonymous but those worthy of mention include: Ajoy Data, Alain Phe, Albert Gasake, Alessia Radice, Alia Hirji, Andreas Stensland, Annelies Ollieuz, Annetta De Vet, Arja Oomkens, Auke Boere, Benecite Giaever, Brenda Sinclair, Carly de Wit, Catharine Russell, Celestin Karamira, Céline Grey, Charles Karangwa, Courtney Blodgett, Daniel Magrizos, Daniel Mcavoy, David Lahl, David Russell, David Williams, Debora Randall, Ed Humphrey, Ellie Dart, Erin Boyd, Estelle Lantin, Fabian McKinnon, Francois Widmer, Fraser Pennie, Geoff Coyne, Georgie Fienberg, Gill Garb, Hannah Matthews, Hatty Barthrop, Hazel Douglas, Ido Verhagen, Janno van der Laan, Jeannetta Craigwell Graham, Jeff Riley, Jemma Hogwood, Jennifer Lentfer, Jessi Smolow, Josep Subirana, Juliette Prodhan, Karen Twining Fooks, Kate Doyle, Kate Mandeville, Katharina Funke Kaiser, Line Loen, Liny Suharlim, Liz Caney, Lucita Lazo, Lynn Dines, Maggie Carroll, Mandy Gardner, Marc de Klerk, Marcella Pasotti, Mariana Infante Villarroel, Marilise Turnbull, Mario Noboa, Matt Bolton, Matt Jackson, Maurice Masozera, Max Perry-Wilson, Mbacke Niang, Mhoira Leng, Michael Brosowski, Milton Funes, Momoko Harada, M. R. Thomas, Mya Gordon, Nansubuga Mubirumusoke, Nicholas Meakin, Niels Hanssens, Olivia Zank, Pam Steele, Patrice Boa, Phil Crosby, Pyt Douma, Richard Labelle, Richard Stuart, Rieke Weel, Robert Nurick, Robyn Kerrison, Samantha Wakefield, Samuel Munderere, Sandra Ondogwu, Sarah Terlouw, Seki Hirano, Shandana Mohmand, Shreela Chakrabarti, Sian Rogers,

Silla Chow, Sive Bresnihan, Solange Baptiste, Sophie McCann, Tara Lyle, Ted Schrader, Thea Lacey, Thurein, Vanessa Baird, Victor Monroy, Weh Yeow, Will Snell, Zach Gross and Zach Warner.

My publishers at Routledge also deserve a very special mention: Andrew Mould for picking up the idea from the start, Sarah Gilkes for the finishing phases and Faye Leerink for the in-between – requiring endless patience, gentle prodding and continuous changing of the publication date as I found yet more people to interview.

The spaces that provided the inspiration and tranquillity required to bring this book together include Altea and Finestrat libraries (Spain), rooftop terraces in Fez Medina (Morocco) and the late-night cool air and red misty dawns of Kigali (Rwanda).

Last but by no means least, numerous friends who supported me in the process and offered endless words of encouragement; special mentions are due to Lucy-Anna Kelly, Jessi Smolow, Michela Fanara and Monique Drinkwater.

ABBREVIATIONS

AAA	Accra Agenda for Action
ADB	Asian Development Bank
AECID	Agencia Española de Cooperacion Internacional y Desarrollo
AFD	Agence Française de Développement
AfDB	African Development Bank
AfP	Alliance for Peacebuilding
AGCI	Agencia de Cooperacion Internacional de Chile
AGI	Africa Governance Initiative
AIDS	Acquired Immuno-Deficiency Syndrome
ALC	Access Livelihoods Consulting
ALNAP	Active Learning Network for Accountability and Performance in Humanitarian Action
ARC	Australian Red Cross
BRICS	Brazil, Russia, India, China and South Africa
C4D	Communication for Development
CAFOD	Catholic Agency for Overseas Development
CAP	Consolidated Appeal Process
CAR	Central African Republic
CBHA	Consortium of British Humanitarian Agencies
CBO	community-based organisation
CCF	Community Capitals Framework
CD	Country Director
CDC	Centre for Disease Control
CDI	Centre for Development Innovation
CDPM	Certified Development Project Manager
CHL	Certification for Humanitarian Logistics
CI	Caritas Internationalis

CIHC	Center for International Humanitarian Cooperation
CILT	Chartered Member of the Institute of Logistics and Transport
CIPS	Chartered Institute of Purchasing and Supply
CIS	Commonwealth of Independent States
CMAM	community management of acute malnutrition
CODEV	Cooperation & Development Centre
CPF	Centre for People's Forestry
CRED	Centre for Research on the Epidemiology of Disasters
CRS	Catholic Relief Services
CSO	civil society organisation
CSR	corporate social responsibility
DAC	Development Assistance Committee (part of OECD)
DANIDA	Danish International Development Agency
DCED	Donor Committee on Enterprise Development
DDR	disarmament, demobilisation and reintegration
DECC	Department for Energy and Climate Change
DFID	Department for International Development
DM&E	design monitoring and evaluation
DOTS	directly observed treatment, short course
DRC	Danish Refugee Council
DRC	Democratic Republic of Congo
DRR	disaster risk reduction
EFA	Education for All
EIA	environmental impact assessment
EKN	Embassy of the Kingdom of the Netherlands
ELRHA	Enhanced Learning and Research for Humanitarian Assistance
eMTCT	elimination of mother to child transmission
EPLO	European Peacebuilding Liaison Office
EU	European Union
EUCORD	European Cooperative for Rural Development
FAO	Food and Agriculture Organization
FFA	Forum on the Future of Aid
FIDH	Fédération Internationale des Ligues des Droits de l'Homme
G8	The Group of Eight (G8) – forum for the governments of eight leading industrialised countries
GAD	Gender and Development
GBV	gender-based violence
GIS	geographic information system
GIZ	German Federal Enterprise for International Cooperation
GNC	Global Nutrition Cluster
GNI	gross national income
GNP	gross national product
GPYE	Global Partnership for Youth Employment
HAP	Humanitarian Accountability Partnership

HDI	Human Development Index
HFH	Habitat for Humanity
HHP	health and hygiene promotion
HIPC	heavily indebted poor countries
HIV	Human Immuno-deficiency Virus
HLDP	Humanitarian Leadership Development Project
HQ	headquarters
HR	human resources
HSK	Harambee Schools Kenya
IBRD	International Bank for Reconstruction and Development
ICRC	International Committee of the Red Cross
ICS	International Citizen Service
ICT	information and communication technology
ICT4D	ICT for development
IDA	International Development Association
IDP	internally displaced people
IDS	Institute of Development Studies
IFAD	International Fund for Agricultural Development
IFC	International Finance Corporation
IFRC	International Confederation of Red Cross and Red Crescent Societies
IGA	income generating activity
IGES	Institute for Global Environmental Studies
IHDI	Inequality-adjusted HDI
IIED	International Institute for Environment and Development
IIEP	International Institute for Educational Planning
ILO	International Labour Organization
IMC	International Medical Corps
IMF	International Monetary Fund
IMRD	International MSc in Rural Development
INEE	Inter-Agency Network for Education in Emergencies
INGO	international non-governmental organisation
IOM	International Organisation for Migration
IPDET	International Program for Development Evaluation Training
IQC	indefinite quantity contract
IRC	International Rescue Committee
ITPC	International Treatment Preparedness Coalition
IUED	Institut Universitaire d'Etudes du Développement
IWMI	International Water Management Institute
IYIP	International Youth Internship Programme
JICA	Japanese International Cooperation Agency
JPO	Junior Professional Officer
KOICA	Korea International Cooperation Agency
KPI	key performance indicator

LAC	Latin America and the Caribbean
LDC	least developed country
LGBT	lesbian, gay, bisexual and transgender
LIC	low income country
LLB	Bachelor's in Law
LLM	Master's in Law
LMIC	lower- and middle-income country
LMS	learning management system
LSDP	Logistics Skills Development Programme
LSHTM	London School of Hygiene and Tropical Medicine
LSTM	Liverpool School of Tropical Medicine
LUMS	Lahore University of Management Sciences
LuxDev	Luxembourg Development
M&E	monitoring and evaluation
M4P	markets for the poor
MANGO	Management Accounting for Non-Governmental Organisations
MBA	Master's in Business Administration
MCC	Millennium Challenge Corporation
MDB	multilateral development bank
MDG	Millennium Development Goal
MENA	Middle East and North Africa
MFI	microfinance institution
MIDA	Migration for Development in Africa
MoH	Ministry of Health
MPA	Master's in Public Administration
MPH	Master's in Public Health
MSF	Médecins sans Frontières
MSMEs	micro, small and medium enterprises
MV	Millennium Village
NABC	Netherlands–African Business Council
NACA	National Agency for the Control of HIV AIDS
NCRE	National Competitive Recruitment Exam
NGO	non-governmental organisation
NHS	National Health Service (UK)
NOHA	Network on Humanitarian Assistance
NRC	Norwegian Refugee Council
NRM	natural resource management
NYU	New York University
OCHA	UN Office for the Coordination of Humanitarian Affairs
ODA	Official Development Assistance
ODI	Overseas Development Institute
OECD	Organisation for Economic Cooperation and Development
OHCHR	Office of the High Commissioner of Human Rights

OIE	World Organisation for Animal Health
OPM	Oxford Policy Management
OVC	orphans and vulnerable children
P4P	Purchase for Progress
PAR	Participatory Action Research
PATHS	Partnership for Transforming Health Systems
PCDN	Peace and Collaborative Development Network
PM4NGOs	Project Management for NGOs
PMI	Project Management Institute
PPP	Public Private Partnership
PRA	participatory rural appraisal
PSD	private sector development
R&R	rest and recuperation
RAPID	Research and Policy in Development
RS	remote sensing
SAP	Structural Adjustment Policy
SGBV	sexual and gender-based violence
SIDA	Swedish International Development Agency
SLA	Sustainable Livelihoods Approach
SME	small and medium enterprise
SOAS	School of African and Oriental Sciences
SSA	sub-Saharan Africa
SSR	security sector reform
SSRRR	Shelter and Settlement Rapid Response and Recovery
STIs	sexually transmitted infections
SUN	Scaling Up Nutrition
SV	Skills Venture
SWAT	Surface water treatment system
THET	Tropical Health Education Trust
ToR	terms of reference
ToT	training of the trainers
UK	United Kingdom
UN	United Nations
UNAIDS	Joint United Nations Programme on HIV and AIDS
UNDP	United Nations Development Programme
UNEP	United Nations Environment Programme
UNESCO	United Nations Educational, Scientific and Cultural Organization
UNFCCC	United Nations Framework Convention on Climate Change
UNFPA	United Nations Population Fund
UNICEF	United Nations Children's Fund
UNIDO	United Nations Industrial Development Organization
UNHCR	United Nations High Commissioner for Refugees

UNOCHA	United Nations Office for the Coordination of Humanitarian Affairs
UNODC	United Nations Office against Drugs and Crime
UNOPS	United Nations Office for Project Services
UNV	United National Volunteers
UNWTO	United Nations World Tourism Organization
VASS	Vietnamese Academy of Social Sciences
VSF	Vétérinaires Sans Frontières
VSLA	Village Savings and Loan Association
WASH	water, sanitation and hygiene
WB	World Bank
WFP	World Food Programme
WHO	World Health Organization
WID	Women in Development
WRI	World Resources Institute

INTRODUCTION

When I started on this project, over three years ago, and engaged Routledge's interest, I didn't quite realise what a task it would be (a friend has dubbed it *War and Peace*). In this time I have profiled 54 different sectors and job roles, interviewed over 100 people and been permanently tuned in to any career news related to the sector.

Many have asked me what triggered the idea for the book. I became interested in the development sector after my first degree, but I combined some years of work and travel before going on to do a general Master's in Development Studies. But even after securing my first job I found career opportunities in the fields of development and humanitarian assistance complex and difficult to navigate. What skills could I market? Where did I want to go? The options seemed vast but opportunities limited.

Around this time I came across a guide to medical careers. I started to envy the set career trajectories of the more conventional careers I had previously shunned. They still required important decisions to be made, but the routes in and up were more clearly defined and documented. There was no equivalent for the development and humanitarian sectors, so I set about writing the book that I felt that I, and the sector, needed.

As I now prepare to submit the manuscript, the feeling is bittersweet. It has been fascinating to engage with so many leaders in the field of development and humanitarian work, some veterans, others still finding their path. I have learnt a lot from their experiences and hope that through this book, you will too. One certainty I have is that, without exception, everyone in this field has a very interesting story to share.

Maïa Gedde
Kigali, February 2014

PART 1

The sector

This first section will help to give you an understanding of what working in the sector involves, where you could work, and who the major employers are. It helps to give those new to the sector some perspective of the field and what opportunities may exist for you within it.

- Chapter 1: Development and humanitarian assistance provides a brief introduction to these two fields, touches on some of the main discourses and historical defining events, and explores the interconnection between them.
- Chapter 2: Is it for you? explores some of the aspects that most often attract people to the sector, and addresses some of the common concerns.
- Chapter 3: Who could you work for? The employers within the field of development and humanitarian assistance are broad; this chapter breaks down the leading organisations into the major sub-sectors.
- Chapter 4: Where could you work? Whether based in a donor or recipient country, jobs primarily follow the aid channels. This chapter helps you to understand where these are concentrated.

PLATE 1 Catholic Relief Services' shelter and settlement technical advisor Seki
Hirano holding a focus group discussion with Tuareg refugees on the
border of Mali and Niger in 2012.

Source: CRS/Seki Hirano

1

DEVELOPMENT AND HUMANITARIAN ASSISTANCE

If you are new to the world of development and humanitarian assistance you probably have many questions. Before embarking on the journey this book will take you through – career opportunities – you need first to understand the basics of the sectors themselves. What actually is development? What is its goal? What is the difference between working in a development and a humanitarian crisis? What might you be more suited to?

This short chapter cannot cover all these issues in detail but aims to give you a very brief introduction to these two very broad fields of work and highlights some of the key issues and trends within them. Appendix 1 offers a reading list of books and resources that can help you to take this journey further.

What is development?

> A developed country is not a place where the poor have cars. It's where the rich use public transportation.
>
> *Gustavo Petro, Mayor of Bogotá*

Development is a complex issue, not always easy to define and thus so difficult to achieve. A principal focus of development is poverty reduction. In an attempt to help gauge the extent of poverty around the world, the World Bank (WB) developed the concept of *absolute poverty* in the 1990s and the international poverty line was anchored at $1.08 dollars per day. This was the minimum income needed to meet a person's basic needs such as food, clothing and shelter by the standards of the world's poorest countries. It is typically measured with respect to a household's average income, not the individual. Thus, the poorest of the poor – the target for many development programmes – are those who subsist on *less than $1 per day*. In 2008 this line was revised to $1.25 dollars a day[1] and *moderate* poverty as less than $2 a day.

Many development programmes aim, either directly or indirectly, to reduce extreme poverty. But the concept of absolute poverty is troublesome as it fails to meet broader quality-of-life issues or address the levels of inequality in society. These criticisms led to the concept of *relative poverty*, which defines poverty in relation to the economic status of other members of the society: people are poor if they fall below prevailing standards of living in a given societal context. By this definition we find a large number of poor people in the so-called *developed* countries.

Both these concepts are, however, largely concerned with income and consumption but poverty can be multidimensional in nature. Social exclusion and lack of access to basic human rights, such as health, shelter, education and sanitation – for whatever reason – are also an issue. Infrastructure and access to such services is also important.

The United Nations Development Programme (UNDP) uses a wide definition of development: to them, it is 'to lead long and healthy lives, to be knowledgeable, to have access to the resources needed for a decent standard of living and to be able to participate in the life of the community'. For others, development is the process of finding creative solutions to chronic problems of hunger, poverty, disease, joblessness and powerlessness (Monterey Institute of International Studies).

The Human Development Index (HDI) is used as a measurement to capture social and economic development by combining indicators of life expectancy, educational attainment and income into a composite human development index. The HDI was first produced in 1990 by the Pakistani economist Mahbub ul Haq and the Indian economist Amartya Sen, and published by the United Nations Development Programme. It started as a single statistic, whereas now it can be disaggregated by indicator. In 2010, the Inequality-adjusted HDI (IHDI) was introduced. In 2012 the country with the highest HDI rank was Norway, and the one with the lowest was Niger.

Grand ambitions, realities and the future of development

Global poverty continues to be a tremendous problem today and no less striking than it was immediately after the Second World War era. Some of the key issues have changed, but as progress is made in one area, new challenges emerge in another. Extreme poverty still affects one in six people worldwide.[2] One criticism is often made that development aid is not impartial, and there are often political agendas hidden behind poverty reduction.

Development projects thus happen in complex and multifaceted situations, where narrow approaches do not work. The challenge is that often development projects are designed with a shallow understanding of the broad and complex contexts in which projects exist. Poverty reduction is not a linear path, and cannot be achieved by climbing up the rungs of a ladder. We have learned that churning out volumes of plans and programmes to accomplish ambitious targets in just three years does not work. We also need to recognise that the *development* of 'rich' countries happened

A BRIEF HISTORY OF DEVELOPMENT

The concept of development has its origins in the industrialization of the nineteenth century, when societies were being reshaped by economic and political changes. But it was only after the end of the Second World War (1939–1945) that the idea of development resurfaced with the meaning we use today. International institutions such as the United Nations (UN), the World Bank, and the International Monetary Fund (IMF) were created to assist Europe in its economic recovery from the war. While Europe rebuilt itself and Europe's colonies started on a path towards independence, the US embraced its new position of global leader and drafted a new foreign policy for the advancement of less economically developed areas. This was a global movement for peace and reconstruction and the start of *globalisation*.

As prevailing theories on the best way to *do* development changed, as did international priorities, distinct *development decades* emerged. A few characteristics of these are:

- 1960s: As post-war Europe recovered, they started to see themselves as part of a movement for change and 'progress', pooling their resources to assist the poor. In 1963 US President Kennedy inaugurated the PeaceCorps scheme to promote social and economic development. African nations had or would imminently gain independence. But the approach adopted was top-down and paternalistic. As developing countries took centre stage in the majority at the UN, it was a decade full of optimism and the 1961 UN plenary meeting declared it the United Nations Development Decade (the first of many). But visionary leaders such as Patrice Lumumba in Zaire (now DRC) faced a lot of meddling from the former colonial powers when these leaders did not serve their interests. A pattern of military coups also started to define African politics. At the same time, in the east, the Asian Tigers of Hong Kong, Singapore, South Korea and Taiwan started their industrialisation and maintained rapid growth rates that by the end of the century placed them in a high income country status.
- 1970s: The landmark 1974 Sixth Special Session of the United Nations General Assembly called for a New International Economic Order to eliminate the unjust trading patterns inherited from colonialism. But at the height of the cold war, the optimism from the 1960s faded when the Organization of the Petroleum Exporting Countries (OPEC) oil price was hiked in the mid 1970s, plunging the world into a recession. Oil-exporting countries had large amounts of extra money which Western banks used to grant huge loans on very relaxed terms to Third World countries, burdening these countries with impossible repayment rates for decades to come. Donors also engaged in large-scale development projects mostly ill-fitted to the local contexts, resulting in failures: these *White Elephants*

of development occurred where misallocation of investment with negative social surplus often benefited local politicians. Import substitution industrialisation was also advocated.

- 1980s. Reagan's trickle-down economics favoured the wealthy and high-income earners and in the 1980s loans were no longer offered as freely as in the 1970s. Following the neoliberalism ideology spearheaded by the 'Washington Consensus' institutions – The World Bank and IMF – loans were now provided on the condition that a country adopt Structural Adjustment Policies (SAPs) to ensure debt repayment and economic restructuring. It required poor countries to reduce spending on things such as health, education and development, while debt repayment and other economic policies were made the priority. Unfortunately SAPs had a negative impact on the social sector and reduced the standard of living of many of their citizens. For example, in health, SAPs affected both the supply of health services (by insisting on cuts in health spending) and the demand for health services (by reducing household income, thus leaving people with less money for health).

- 1990s: This decade saw an agenda for change brought about by bad governance and authoritarian regimes. The emphasis was on good governance and democracy, and participative approaches to development and the environment were on the agenda. *Local ownership, sustainability* and *bottom-up approach* became buzz-words. But change was slow: despite the fact that the 1990s saw a drop from 30 per cent to 23 per cent in the number of people globally living on less than a dollar a day – mostly due to progress seen in China and India – this is commonly known as the *lost decade of development*. While the US and much of Europe were booming, more than 50 countries suffered falling living standards. The UN attributed much of the decline to the spread of HIV/AIDS, which lowered life expectancies, and to a collapse in incomes, particularly in the commonwealth of independent states. Many countries were spending more in servicing debts to the World Bank and the IMF than they were spending on health and education.

- 2000s: The start of the new century saw a burst of enthusiasm to end global poverty. At the Millennium Summit of the United Nations in 2000, eight international development goals – the Millennium Development Goals (MDGs) – were established and agreed by all 189 United Nations member states with a target date of 2015. The eight goals were to:

1   eradicate extreme poverty and hunger;
2   achieve universal primary education;
3   promote gender equality and empower women;
4   reduce child mortality rates;
5   improve maternal health;

6   combat HIV/AIDS, malaria, and other diseases;
7   ensure environmental sustainability;
8   develop a global partnership for development.

Not everyone agreed with the MDGs of course, as they felt certain areas – such as human rights and disability – were distinctly lacking. But what they did succeed in was to unite the development community and bring about more collaboration as agreed during the Paris Declaration. 9/11 also created a consensus that poverty was the world's problem, but the UN urged the West to abandon the one-size-fits-all liberalisation agenda foisted on poor countries. The biggest ever anti-poverty movement came together under the banner of MAKEPOVERTYHISTORY. Together with the Drop the Debt Campaign and others, they lobbied the G8 at Gleneagles in 2005, demanding that rich countries stop blocking a global trade system. 'Trade not aid' became the slogan. The G8 Finance Ministers also agreed to provide funds to the World Bank, IMF and the African Development Bank (AfDB) to cancel $40 to $55 billion in debt owed by members of the heavily indebted poor countries (HIPC) allowing them to redirect resources to help accelerate the progress towards the MDGs which was a huge achievement. Rich countries also committed to scale up their contribution to 0.7 per cent of their gross national product (GNP) towards official development assistance (ODA) by 2015, although such a pledge was also made back in 1970[3] and not upheld. Following the massive media attention issues of poverty and development received, more and more people started to plan a career in the sector, with Master's programmes in International Development becoming increasingly popular in universities around the world.

• 2010s: The 2008 global economic crisis that originated in developed countries only started to have a visible effect at the start of this decade. Aid budgets were cut with stand-alone international development departments such as the Canadian International Development Agency (CIDA) – a 45-year-old federal agency – which was merged into the Department of Foreign Affairs and International Trade in 2013.[4] This emphasised a conservative move to keep trade policy as a key part of their foreign affairs agenda. Others, such as Australia's AusAID followed suit. Budget cuts were accompanied by tighter donor demands around effectiveness and efficiency. With the Arab Spring, youth unemployment hit the world agenda as it impacted on economic growth, security and stability. Climate change and food security were also high on the agenda. The global development agenda for the post-2015 period was initiated in 2010 and has sustainable development at its centre. The website post2015.org brings together this thinking.

under very different circumstances to what 'poor' countries currently face, so neither the process nor the end product can nor should serve as a model. Creative, innovative, multifaceted solutions are needed.

Two bodies of thought now predominate. One focuses in on the reality that there is no quick fix solution: more money and longer term investments are needed to bring about development. The other one asks what is there to show for the hundreds of billions of dollars of international development assistance that have been invested, since children still die for lack of 12-cent medicines to prevent malaria? Do we really need to invest more in this bottomless pit? This discontentment about the effectiveness of aid, together with the most recent global financial crisis, has seen aid budgets significantly cut over the last few years.

While many challenges do remain, it is important not to forget some of the successes that have been made. Table 1.1 highlights some of these successes in the last 50 years.

As a sector, the development industry is relatively new and is professionalising. It has evolved significantly from its rather ambitious and naive origins over six decades ago to a mature and professional industry today. But most would agree that significant reforms are still needed to start producing real, lasting change. The good news is the sector is anything but stagnant: those who have worked in the field for some years well know the dynamic nature and rapidly changing priorities. As a development professional you will constantly need to be learning, adapting and innovating and, most likely, changing your focus several times in your career. This in itself may be one of the challenges – the constant search for the silver bullet, the solution to end poverty. In light of this, what does it mean for the person carving a career in this field? What are the trends we will see over the coming years?

In light of these changes some of the trends we will see in future might be reduced donor budgets and increased emphasis on private sector development, public private partnerships, entrepreneurship and job creation. There is a real emphasis on what really works, grounding development projects in the knowledge base. Impact evaluations with rigorous methodology will be in greater demand. The post-MDG agenda has committed to 'End Poverty' as the first goal and 'Leave no one behind' as the first of five 'transformative shifts'. Despite this, the social impact of economic downturns and crises will hit the most vulnerable hardest and humanitarian capacity will be essential to prevent these crises resulting in long-term poverty.

Humanitarian assistance

While development concerns itself with poverty reduction, humanitarian assistance is designed to save lives, alleviate suffering and maintain and protect human dignity during and in the aftermath of emergencies.[5] There are of course several links between extreme poverty and conflict, insecurity, vulnerability and humanitarian

TABLE 1.1 Selection of achievements and remaining challenges

Sector	Key achievements	Remaining challenges . . .
Health	• In the past 20 years the number of the world's chronically undernourished has been reduced by 50% • In the past 50 years infant and child death rates in the developing world have been reduced by 50% (Source: USAID) • Smallpox was eradicated worldwide in 1979	• About 29,000 children under the age of five – 21 each minute – die every day, mainly from preventable causes (70% of these deaths are attributable to just six causes: diarrhoea, malaria, neonatal infection, pneumonia, preterm delivery or lack of oxygen at birth, and occur mainly in the developing world) (Source: UNICEF) • Ten countries in sub-Saharan Africa have less than three doctors per 100,000 people. The global average is 146 per 100,000 (Source: WHO 2006)
Food	• The Green Revolution and improvement in farming methods has meant that food yields per hectare are higher today that at any other point in human history	• 842 million people in the world do not have enough to eat (although this number has fallen by 17% since 1990) • In developing countries, rising food prices form a major threat to food security, particularly because people spend 50–80% of their income on food (Source: www.un.org/waterforlife decade/food_security.shtml) • One in four of the world's children are stunted from lack of food (Source: www.wfp.org/hunger/stats)
Democracy and governance	• There were 58 democratic nations in 1980. By 1995, this number had jumped to 115 nations (USAID)	• Less than 1% of what the world spent every year on weapons was needed to put every child into school by the year 2000 and yet it didn't happen • Corruption is still an issue that hinders the development of many countries

continued . . .

TABLE 1.1 Continued

Sector	Key achievements	Remaining challenges . . .
Economic growth and financial independence	• Some African countries now have an average growth rate of 7% – much higher than present-day Europe – and similar to the growth levels that propelled the Asian Tigers into high income countries	• Inequality is rising. The poorest 40% of the world's population accounts for 5% of global income. The richest 20% accounts for three-quarters of world income • Nearly 40 per cent of the world's unemployed – about 81 million – are between 15 and 24 years of age. More youth are poor and underemployed than ever before posing a risk to a country's stability
Education	• Literacy rates are up 33% worldwide in the last 25 years • Primary school enrolment has tripled in that same period (USAID)	• Nearly a billion people entered the twenty-first century unable to read a book or sign their names (UNICEF) • While MDG 2 has been met with 90% enrolment in primary education, dropout rates remain high, quality is often poor and 57 million children remain out of school (Source: www.un.org/millenniumgoals/education.shtml)
Environment	• Deforestation in developing countries has slowed down (but still remains higher as 2.5 billion people in Africa and Asia rely on biomass fuel)	• Some 1.1 billion people in developing countries have inadequate access to water, and 2.6 billion lack basic sanitation provisions • The population of the 49 least developed countries is growing nearly twice as fast as the rest of the developing world

ENDING POVERTY: A REALITY IN OUR GENERATION?

In the 2005 bestseller, *The End of Poverty*, economist Jeffrey Sachs argued that with proper planning and funding, extreme poverty could be wiped off the Earth by 2025. In 2005 he set out to achieve what others – and millions in aid – had failed at. Sachs, the director of the Earth Institute at Columbia University and special advisor to the UN Secretary-General on the MDGs, teamed up with Raymond Chambers, a pioneer of private equity investing and a passionate philanthropist, to establish Millennium Promise through the Millennium Villages (MV). It was an interesting idea: take a village, address the root causes of extreme poverty, taking a holistic, community-led approach to sustainable development by uniting science, business, civil society and government, and eradicate poverty. Each village representative was armed with a 147-page, *Millennium Villages Handbook*, written by 30 academics, which detailed, intervention-by-intervention, how to reorder village life and thereby eradicate poverty. Between 2005 and 2015 the aim was to show how these villages could provide a scaleable sustainable model for development with the promise of eradicating extreme poverty, hunger and preventable disease. Public, private and non-profit partners showed their commitment to ending extreme poverty by supporting Millennium Promise. Fifteen villages in Africa were selected. Where others have failed has Sachs succeeded? Now nearing their end in 2015, progress has definitely been made (but so too it has outside the selected villages) but the anticipated levels of success have not been met. Sadly, it seems that the solution to ending poverty cannot be found in a 147-page manual.

www.millenniumvillages.org

Jeffrey Sachs (2005) *The End of Poverty: How We Can Make it Happen in Our Lifetime*, New York: Penguin.

Nina Munk (2013) *The Idealist: Jeffrey Sachs and the Quest to End Poverty*, New York: Doubleday Books.

crises. The poorest countries invest the least in disaster preparedness, so if a disaster hits, it is usually beyond the capacity of national authorities alone, so the international community will be mobilised to respond. These countries also have high population densities in poorly planned urban areas, so the number of people affected is large. And there is a close link between poverty, unemployment and stability.

Humanitarian assistance tries to ensure the best possible outcomes for people affected by disasters and crises. The characteristics that distinguish it from other forms of foreign assistance and development aid are:

- it is intended to be governed by the principles of humanity, neutrality, impartiality and independence;
- it is intended to be 'short term' in nature and provide for activities in the 'immediate aftermath' of a disaster.

Emergency response deals with the immediate aftermath of emergencies. This can include the provision of material relief assistance and services (shelter, water, medicines) emergency food aid (short-term distribution and supplementary feeding programmes) relief coordination, protection and support services (coordination, logistics and communications).[6] However, many emergencies result in prolonged vulnerability and some organisations stay on to invest in long-term reconstruction efforts – a middle ground between development and disaster response.

Today, humanitarian funding comes through the UN Consolidated Appeal Process (CAP), although the International Confederation of Red Cross and Red Crescent Societies (IFRC) and other organisations run their own appeals too. It is a process that brings aid organisations together to jointly plan, coordinate, implement and monitor their response to natural disasters and complex emergencies. It also allows them to appeal for funds cohesively, not competitively, and based on a consolidated humanitarian action plan.[7]

Government donors give the largest amount of humanitarian assistance, on average accounting for over 70 per cent of the international humanitarian response since 2007. Private donors are thought to have provided more than a quarter of all humanitarian assistance over the past five years, largely in the form of voluntary contributions from the public to non-governmental organisations (NGOs).

The number of inter- and intra-state conflicts has steadily been reduced over the past 20 years. Nevertheless, tens of millions of people are still affected by protracted conflicts and some new conflicts, and the number of people internally displaced by armed conflict around the world is increasing – from about 17 million in 1997 to an estimated 26.4 million people at the end of 2011.[8] The effects of armed conflicts are felt most strongly by vulnerable communities in countries such as Afghanistan, the Democratic Republic of Congo, Somalia and South Sudan.

At the same time, the frequency and intensity of natural disasters is increasing in magnitude and the impact is felt most acutely in the developing world with Africa and Asia the hardest hit. Over the last 30 years, people on these continents have made up approximately 88 per cent of the total number of reported deaths, and 96 per cent of the people affected by natural disasters.[9]

The humanitarian landscape is evolving, with global challenges – such as climate change, population growth,[10] rapid and unplanned urbanization, and food and water insecurity – leaving more and more people at risk. At the same time, new types of organisations are becoming increasingly involved in emergency response, from telecommunications companies to diaspora groups.[11] Humanitarian needs are expected to continue to rise in the coming decades.

As the world becomes more industrialised, we are likely to see an increase in technological catastrophes as a result of natural hazards affecting areas such as oil

SOME KEY EVENTS IN THE EVOLUTION OF HUMANITARIAN ASSISTANCE

Before the Second World War, only a small number of organisations were dedicated to humanitarian assistance, and the International Committee of the Red Cross (ICRC, founded in 1863 with the objective of ensuring protection and assistance for victims of armed conflict and strife) led the way. Humanitarian efforts were concentrated in the immediate post-Second World War Europe. CARE USA, was founded in 1946 to provide food relief in the form of care-packages to war-devastated Europe.

Events unfolding after the Biafran war in 1968 helped to shape a humanitarianism that we know today. The Eastern part of Nigeria had declared independence, calling their new state Biafra. In response, the Nigerian army attacked the rebel government. Things went very badly for the Biafrans, but no one in the West cared. The British government continued to happily sell arms to the Nigerians. The ICRC was on the ground, but unable to speak out about the events they witnessed, due to its stance of impartiality and neutrality. This brought the criticism that the Red Cross was complicit with the Nigerian Government and their mass killing and starving of between one and three million Biafran people.

The Biafran government did a strange thing, and hired a PR firm in Switzerland to help raise awareness of the atrocities happening in Biafra. They sent daily mail shots and updates out to every leading newspaper and MP in the UK. People started speaking out. Pictures of starving children caught up in the genocide started to appear in the newspapers. People mobilised and this was the start of a humanitarian appeal as we know it.

A young French doctor who was working in Biafra for the Red Cross wouldn't accept remaining silent. So he went about changing the humanitarian landscape and set up Médecins sans Frontières (MSF) in 1971. While still bound by principles of neutrality and impartiality, MSF considers one of its functions to be speaking out on human rights abuses, drawing attention to cases of human rights violations that MSF considers under-reported, and on occasions taking a stronger stand and denouncing egregious violations. Unlike the Red Cross, MSF is also willing to enter war-torn areas without the permission of authorities.

This was the start of a humanitarianism that had the power to right wrongs around the world rather than just alleviate them. This also brought about the idea of 'humanitarian intervention' which was behind the decision to attack Iraq, Libya, and is one of the central beliefs of our age. It divides people. Some see it as a noble, disinterested use of Western power. Others see it as a smokescreen for a latter-day liberal imperialism. The early 1990s saw some disastrous humanitarian engagements in Somalia, Rwanda and Bosnia.[12]

The 2004 humanitarian relief efforts in Sri Lanka post-tsunami drew attention to the need for greater coordination and professionalisation, and the need to impose standards within the humanitarian community. As aid poured into the affected area following the extensive media coverage of the disaster, so too did individuals, small and big relief organisations. Chaos ensued as individuals and organisations tried to get rid of their money in the most effective way – creating competition among agencies which in some cases actually hindered their humanitarian efforts. Small NGOs and individuals with little humanitarian experience also flocked in, frequently undercutting standards. One example was in housing where smaller NGOs built unsafe temporary housing, without any consideration of the future development of the area.

Between 2005 and 2010 the humanitarian sector reformed, with a greater emphasis on coordination and quality standards – a challenge, as in an emergency there is usually very little time for this.[13] The UN has spearheaded a cluster approach, with agencies grouped according to their expertise such as water and sanitation, nutrition, child protection or health. Each cluster is led by a UN body, such as the World Food Programme, United Nations Children's Fund (UNICEF) or the World Health Organization. OCHA is the UN Office for the Coordination of Humanitarian Affairs.

Today humanitarian relief is governed by a series of best practice standards that recommend how organisations should relate to one another and also work with their staff. The Humanitarian Accountability Partnership (HAP) International has introduced one such standard. The voluntary Code of Conduct of the Red Cross and Red Crescent movement and NGOs in Disaster Relief is another. The Sphere Project aims to uphold the standards of the global community and the People in Aid Code of Good Practice related to human resource (HR) management.

pipelines or dams. The humanitarian community may therefore have to respond more frequently to technological and environmental disasters. But still over half of humanitarian assistance is going to long-term recipients where disasters reoccur and people are caught up in a vicious cycle of chronic poverty and crisis.

Future trends show that there will be greater assistance from domestic governments and more spending from non-DAC donors such as Turkey. There is an emphasis and greater need to develop the skills of local aid workers, to build local and national capacities to prevent and mitigate future humanitarian crises. One of the challenges will be around compliance with humanitarian principles, law and access to victims. But digital connectivity is also transforming how we respond to emergencies, as was demonstrated in the Digital Humanitarian Network to Typhoon Haiyan in the Philippines.

The community is also increasingly looking towards disaster risk reduction (DRR) as a way to mitigate the impact of disasters: prevention is always better than cure. Unfortunately, relatively little is yet invested here. The emphasis here is also on developing the resilience of the community, that is:

> the ability of a system, community or society exposed to hazards to resist, absorb, accommodate and recover from the effects of a hazard in a timely and efficient manner, including through the preservation and restoration of its essential basic structures and functions.[14]

Are you more suited to work in a humanitarian or development context?

Working in development and emergency response are quite distinct. Each requires different skills, personality traits, values, outlooks and lifestyles, with many people making a conscious decision to work in one field or another, although many people successfully make the transition between the two.

> I work on food security issues and have done so in both humanitarian and emergency contexts. But the work in both these environments is very different. In humanitarian contexts you are likely to see almost immediately the impact of your work but it is short lived. With development work you have to invest significant time before seeing results, but the focus is on long-term change.
>
> *Charles (Burundi)*

If you are starting out in your career in the field, it is important to get an idea about where your interests lie in order to better plan your entry into the sector. Table 1.2 helps to highlight some of the characteristics and differences between the two extremes of humanitarian and development work. There is of course a middle ground, longer-term reconstruction work, not included here.

Now that you have a broader understanding of the fields of development and humanitarian assistance, you are encouraged to read further about these areas.

Appendix 1 offers some key reading lists, including books, journals and blogs. Chapter 2 will take you on to explore some of the highs and lows of working in the sectors.

> I worked on a social housing project in Honduras after hurricane Mitch destroyed many homes. International NGOs flocked in with support. Many faith-based organisations offered *free* houses. My organisation on the other hand, offered smaller houses, purchased with a loan. I was in charge of this project, with a target to build 5000 houses. You can imagine what a hard time I had, convincing people to take one of our houses instead of a free one.
>
> There were several occasions where I had made an arrangement with the community to start building our houses, and the next day another organisation

TABLE 1.2 Are you most suited to development or humanitarian work?

Development	*Humanitarian assistance*
Job characteristics	
Longer term contracts common (6–36 months or permanent) – unless you are working as a consultant	Short-term missions (1–6 months), unless you are working in longer term reconstruction efforts
Relatively stable work and living environment	Likely to be working in insecure zones with heavy security restrictions
Likely to have a heavy workload but 8–10 hour days possible	Fast-paced and 12-hour days not uncommon – often in a high pressure environment (especially in the immediate aftermath of a conflict or natural disaster)
Results not always immediately visible, working towards longer term objectives	More likely to see immediate results of one's work
Normal holiday allowance	Regular rest and recuperation (R&R) breaks and longer gaps in between missions
May be based in HQ or in the field	Likely to be in the field on missions, unless you have a coordination and management role at HQ
Many positions allow family to accompany	Mostly unaccompanied positions except in more stable environments
Often working on coordination and management of projects, often one step removed from local populations	Working closely with local populations
Potential areas of work: farming and agribusiness, gender, microfinance, reproductive health, governance, rural development . . .	*Potential areas of work*: emergency health care, logistics (e.g. food aid), public health, protection, displacement and refugee camp management . . . (see Part 4)
Personal characteristics	
Adapts well to new environments and cultures	Copes well in stressful and insecure environments with ethical and moral challenges
Ability to understand and respect different points of view, not a 'my way is the right way' attitude	Level headed in face of the social and political complexities of humanitarian emergencies
Ability to see the big picture and form strategic partnerships (networking is a key skill)	Team player, flexible and easy-going in relation to personal need
Ability to see the big picture and the long-term impact. Adapts vision and sense of fulfilment with daily work	Knows personal limits to maintain health and avoid burnout

Note: For professional competencies for Humanitarian Assistance see Appendix 3, Core Humanitarian Competencies Framework.

came in offering their free houses instead. In order to implement the project, we had to reach out to communities beyond the original geographic area of the project, targeting areas that the free housing had not reached, so we managed to reach our target. We were aiming for a repayment rate of 30 per cent, which would then be invested back into a credit scheme to enable people to expand their houses.

A few years later I returned to see how things had gone. The repayment rate was 98 per cent. Much higher than we were expecting. I went to visit some communities, and people had expanded their houses, they were sending their children to school, the communities were thriving. I also visited the communities who had received the free houses and spoke to one of the mayors. He told me he had a big problem. People expected everything for free. They refused to pay for even basic services: water, community charges. People hadn't invested in their houses, some were empty, collapsing.

To me this underpins what development is and isn't. It isn't charity. By giving hand-outs you aren't helping anyone to develop. You need to see poor people with equal eyes. You are not developing them if you are spoiling them. You need to empower them.

Milton (Honduras)

Notes

1 Dollar a Day revisited, Development Research Group, World Bank. Accessed on www-wds.worldbank.org/external/default/WDSContentServer/IW3P/IB/2008/09/02/000158349_20080902095754/Rendered/PDF/wps4620.pdf in January 2014
2 CIDA Closed: International Aid Agency Merged With Foreign Affairs www.huffingtonpost.ca/2013/03/21/cida-closed-budget-2013_n_2926517.html
3 www.unmillenniumproject.org/resources/fastfacts_e.htm
4 Millennium Project. The 0.7 per cent target: an in-depth look www.unmillenniumproject.org/press/07.htm
5 Global Humanitarian Assistance Report 2013, available from www.globalhumanitarianassistance.org/wp-content/uploads/2013/07/GHA-Report-2013.pdf
6 Ibid.
7 www.unocha.org/cap/about-the-cap/about-process
8 www.unhcr.org/4fd6f87f9.pdf
9 Policy and Studies Series 2012, *Coordination to Save lives and History of Emerging Challenges.* http://reliefweb.int/sites/reliefweb.int/files/resources/OCHA%20Coordination%20Study%20(Advance%20Version)%20low%20resolution.pdf
10 The world's population was close to two billion in 1945, hit six billion in 1999, surpassed seven billion during 2011 and is expected to reach 8.4 billion by 2025.
11 OCHA, first humanitarian symposium looks to the future of aid, December 2012 accessed at www.unocha.org/top-stories/all-stories/ocha%E2%80%99s-first-humanitarian-symposium-looks-future-aid
12 GOODIES AND BADDIES. Adam Curtis www.bbc.co.uk/blogs/blogadamcurtis/posts/goodies_and_baddies?comments_page=2
13 Coordination to Save Lives, History and Emerging Challenges 2012. Policy and Research Paper. OCHA https://docs.unocha.org/sites/dms/Documents/Coordination%20to%20Save%20Lives%20History%20and%20Emerging%20Challenges.pdf
14 International Strategy for Disaster Reduction (ISDR) 2009.

2

IS IT FOR YOU?

Are you looking for opportunities for an international and diverse career, working with people from different cultures, overcoming challenges, collaborating with local communities, lobbying at an international level or planning the logistics of an emergency food aid shipment? A career in international development or humanitarian assistance could offer all of this. And your ultimate target will always be to improve the lives of some of the world s poorest people.

The possibilities for a fulfilling career in these dynamic and diverse sectors are many. But is it for you? Along with the highs come the inevitable lows. There are many complex and contradictory issues that may lead to disillusionment. This chapter serves to give you a snapshot of some of the reasons why people are attracted to work in the sector along with some of the downsides.

At the outset it is important to understand your different motivations for working in the field of development or humanitarian assistance. Is it the travel, the challenges, the adrenaline, the desire to help people or the unconventional career? Being honest about what attracts you to the work will allow you to understand whether it is really for you – or whether an alternative field might be a better fit.

What attracts you to the sector?

A worthwhile career

The first thing many people talk about is job satisfaction. Working to improve the lives of the poorest is at the centre of what we do.

> Working in humanitarian logistics has given me the opportunity to bring life-saving supplies and hope to victims of both human-made and natural disasters. I derive much personal satisfaction from knowing that my contribution saves lives and helps restore dignity and hope to the victims of disasters. It drives me to challenge myself and strive to do my best each and every day.
>
> *Pamela (Kenya)*

Especially those who have made the transition to the not-for-profit sector talk about how they feel personally much more fulfilled with the cause they are working towards: instead of making money for the corporation they are helping people improve their lives (if working in development) or directly saving lives (if in the immediate aftermath of an emergency).

> As a construction engineer in Hong Kong I project managed the construction of tall buildings, mostly office blocks. In Sri Lanka I oversaw the rebuilding of housing by people affected by the 2005 tsunami. I found this work much more challenging and fulfilling.
>
> *Silla (Hong Kong)*

For some, the vast inequalities that exist around the world are a moral imperative, for others there is also a religious component. Many faith-based development organisations work in this field.

> Being a Christian, it is important for me to work for an organisation whose mission I fully agree with and share. It makes the work I do feel even more meaningful.
>
> *John (Canada)*

But don't be under the illusion that you will be 'saving the world'. For those starting with high hopes, the bubble of idealism quickly bursts when they experience the realities on the ground, some of the challenges, inefficiencies and complexities of the situation and the slow process that is change. Vast amounts of money might be pumped into a project with little or no long-term visible impact once the funding ends. You are a small cog in an extremely complex machine – where no magic replicable formula exists. Unless you work in emergency contexts, directly saving lives, it can be difficult to see the direct impact of your work. Social change is about evolution not revolution.

> It may sound stereotypical, but my main motivation for working in the sector is to try to make the world a better place. Personally I wanted a career that was meaningful, challenging, worthwhile and of course an opportunity to travel. I go through various phases of disillusionment, but ultimately it offers me all of that.
>
> *Hannah (Spain)*

The work

> I love the challenge and the complexities that come with working in this sector, I learn interesting things every day. I enjoy working directly with communities, local and national governments. I work in shelter and come up with solutions together with the community, the team and peer agencies. Things are never

as they look from the outside. Planning a large-scale intervention that will improve the lives of the people forever is work I would never give up.

Seki (Japan)

Your day-to-day work will very much depend on your organisation and your role, area of specialism and the capacity in which you are working. One thing is for sure, it is likely to be diverse and dynamic. If you work in a technical area, you'll probably find your work is remarkably similar to the work you were doing back at home, just in a different, often more complex and multidimensional environment. Making progress involves tackling a lot of different problems.

Many people have the impression that *working in development* means being out *in the field*, building a school, engaging with the community leaders to decide the location of a well. Early on in your career you may have the opportunity to do hands-on work, but as an international staff member (often commanding a higher salary than local staff), working in a country that is not your own – with all the cultural and language barriers that go with that – your work will more often be at higher level, contributing skills that are not available locally.

> Working 'in the field' is not as exciting as it sounds. The reality is that you are always more glued to your desk than your job description states. You do spend a significant amount of your time at meetings and outside events, and if you are lucky a quarter of your time may actually be spent travelling out to projects in the field.
>
> *Janno (Netherlands)*

On the plus side, the multidimensional aspects of development will often give you chances to innovate, challenge yourself and never be bored.

> Working in development you need to be prepared to multi-task as there are so many different dimensions to the work. Development gives you the opportunity to learn much more than any academic programme, but you must have the willingness to learn. I started my career in development as a Housing Manager (I'm a Civil Engineer and an Urban Planner) but today, 12 years later, I'm leading a community-based programme in Rwanda for orphans and vulnerable children (OVC) affected or infected by HIV and AIDS.
>
> *Milton (Honduras)*

The people

It takes a certain type of person to work in these fields. They have to be not only good at their job, but also fit in with the organisation, have understanding and be able to work within their context. Flexibility, adaptability and cultural sensitivity are qualities that most recruiters are looking for in addition to technical competencies. The people you meet will often be personable, open-minded and interesting – although you may occasionally meet the odd one, and wonder how they got recruited. The success of a development project ultimately depends on the ability of people to work together, building on their strengths and sharing knowledge in joint operations and projects. This interaction is what can make development management so exciting.

> My job gives me the opportunity to meet and work with a hugely diverse range of people. As the director of an orphanage trying to achieve self-sufficiency on a very small budget I need to engage international volunteers. So far we have had NASA engineers building fuel efficient stoves, teachers without borders, journalists writing about us, food processing experts helping us to develop fruit drying techniques for export, architects to design an eco-lodge. On the ground we work closely with the local community, leaders and government, and collaborate with other organisations. In no other job in the world would I get to interact with such a diverse range of people in one place.
>
> *Victor (Guatemala)*

Your colleagues will also have an important role in your life, especially in the humanitarian environment, where tight security restrictions will not allow you to go outside your workplace or accommodation. This creates close and intense friendships and will give you the opportunity to work in multicultural teams and interact with a huge range of people, many of whom will hold similar values and outlooks to yourself. The image of the humanitarian worker might be an adrenaline-charged junkie, but you will meet a whole variety of inspirational people who have taken an alternative career path and possible lifestyle. Although, as everything, there are always two sides to the story.

> As someone who has lived in a developing country for a long time, but not worked in the sector myself, I see a certain arrogance among the development 'expats'. The young people who come out on volunteer placements are full of idealism, proud that they are doing something very worthy, and are going to save the world, with those who have been in the sector for a while, moving from country to country, becoming quite insular. Making friends with locals is hard, so they prefer to stick within their expat circles. They get together to criticise the local culture, with an attitude of superiority, instead of seeing themselves as equals.
>
> *Anonymous*

Travel and lifestyle

Opportunities to travel with your career to interesting and sometimes difficult places are one of the main things that attract people to the sectors. Some jobs might involve frequent short trips; others might take you to live in a country for an extended period of time. Some jobs may be frustratingly static. Travelling can also have its challenges. Relocating after every two or three years can eventually become exhausting, leaving old friends, making new ones. Friendship circles are more international and you lose roots at home. You might miss weddings, births, funerals and general family life. Your life will diverge significantly from old friends', making it difficult for them to understand what you do and find common threads again. If you have not done much travelling, the idea of living overseas can be daunting. Will you make friends and survive in a different culture without your home comforts? If you don't know the answer to this the best way is to try it out by doing some volunteer work or aim for a headquarter (HQ) post.

> I'm based in Nairobi and am a regional Monitoring and Evaluation (M&E) Manager for a large international NGO. My job requires frequent international and regional travel including Uganda, Tanzania, Côte d'Ivoire, Malawi, Ghana and Ethiopia. A job that requires a lot of travel sounds really glamorous until you are actually doing it. I spend most of my time in hotels and conference rooms, not necessarily seeing that much of the country I am visiting. It can be quite difficult socially, especially as an expat if you are constantly on the road and it can be lonely and tiring. Also, getting your routine work in around your overseas travel means late nights in the office.
>
> *Catharine (UK)*

But if you are posted overseas in a nice country, your life can be much more comfortable than it was back home: a larger house, a cleaner, a nanny for your children and being part of the wealthier strata of society, which you may not have been back at home.

> Life here with my two small children is good. We have a lovely live-in nanny making it easier to socialise, nice school, nice weather year around, and a cheaper life with holidays in the national parks. The longer I spend away from my home country, the stranger it seems to go back. In fact I start wondering, why would I want to?
>
> *Anonymous*

What are your concerns?

Competition to get into the sector

> Patience and commitment are the two words to keep in mind if you want a career in international development. There is no set, or easy, route into this sector.
>
> Liz Ford, 'So you want to work in . . . International Development', *The Guardian* Saturday 19 January 2008

All of the perks of the sector, the growing spotlight on development and human-itarianism, greater global awareness and recent disillusionment with the world economic order, have made this an attractive and enviable area to work in. The result? Increasing competition for jobs.

Without doubt, the development sector has become one of the most competitive to get into. It is not uncommon for there to be more than 200 applications, all from perfectly qualified people who fit the person specification, for a single job. Ten years ago or so you might have stepped into a job with just a bit of passion and some international experience; today a Master's has almost become an industry standard, but is by no means a guaranteed entry ticket. You'll also have to have hard *in-demand* skills, as well as substantial international experience and even then you will be up against several equally qualified candidates.

> I had a Master's in Development, I had international volunteer experience, I went to a good university and had a good degree, and I spoke three languages. But I applied for over 50 jobs and didn't even get a single interview. It was very disheartening but I had to persist. Eventually, a low-paid internship led to a temporary (paid) position, which eventually became a full-time job.
>
> *Mary (France)*

Many of those who recently entered the sector and are in good jobs today have tales of years of hardship, holding on to two jobs to fund volunteer international work overseas, studying, churning out job applications and several months of unpaid work. It may be tough going, but eventually persistence does pay off. Part 2 of this book guides you through the process of breaking into the sector. An amount of serendipity and luck are required, but good career planning and gaining hard skills are also important.

Skills required

> Being good-hearted is not good enough. Indeed, if your passion is giving, this might *not* be the right field to break into. It's very important to distinguish between philanthropy and development. To work in development you need to be a certain person, have certain skills to offer and the capacity to transmit/teach them to others. Remember the Chinese Proverb: *Give a man a fish and you feed him for a day. Teach a man to fish and you feed him for a lifetime.*
>
> *Milton (Honduras)*

The sector has definitely professionalised in the last few decades with funding coming under greater scrutiny and its effectiveness being questioned. As countries build their local capacity there is less space for international staff.

Once you are in and get known for producing good quality work, it is likely that new jobs will be offered to you without the need for filling in too many job applications.

> I haven't really had to apply for a job in several years. In fact most people I know get offered work through word of mouth. Once you are in the sector and people know your work it becomes easier to be recruited through recommendations.
>
> *Anonymous*

Outside perception of the sector

> I chose to focus on humanitarian work (which is defined by impartiality) as I didn't like all the hidden agendas of the donors in the development sector.
>
> *Anonymous*

For all its failings the development sector can be viewed quite negatively. From the outside the perception of people who are 'amateurs trying to do good in the world' is not uncommon for NGO workers. As a field once defined by goodwill and volunteering and the frequent bad press of development's failings, it is not hard to see why this is. With few structured career paths, it is difficult to clearly see how (or where) you can get to the top of your career.

The overall perception of the multilateral and bilateral agencies has also gone down, with people who are highly paid but ineffectual in the system. We cannot know whether it has done more harm than good, but this is a common argument that aid sceptics use.

> I came from the private sector to work for an NGO and although I enjoyed the work, I didn't want to stay for more than two years in the field. I was worried that future private sector employers would not view me as serious if I stayed much longer, and I was worried about lack of rigour that I experienced myself.
>
> *Anonymous*

Personal relationships

Personal relationships can be a challenge. In the highly stressful environment of emergency relief, you will turn to others for support, which can create strong bonds between people that don't necessarily make much sense outside the context. Many development jobs will allow your family to accompany you, but in acute emergency situations this is unlikely to be possible, although in reconstruction efforts it is more feasible. But with two people juggling their career and frequent relocation, the divorce rate for those working within the sector is also very high. The best you can do is choose an equally adventurous or tolerant spouse who has a flexible and mobile career.

> I just wanted to travel. That was my sole motivation for choosing my degree (in water management) as my university ran yearly field trips. I then worked for many years in Burkina Faso, Cameroon and Mali. So then I was no longer

travelling, it becomes normal life, you make it your home and you form bonds there. You need to be prepared for this. My wife is now Cameroonian and we have three children. We did live back in Europe for a while but the transition was hard for her so we settled back in Africa.

Anonymous

Financial remuneration

Gaining the skills, knowledge and experience to demonstrate your commitment and contribution to the sector will take time and possibly monetary investment on your behalf. You may have to spend a period doing unpaid or low paid work, as well as an advanced degree. Will it pay off in the end?

There are huge salary discrepancies within the sector. One person might be working for an international organisation receiving a very generous international package with many perks (housing, schooling, relocation, settling-in allowance . . .), while an equally qualified person might be working for a smaller organisation receiving a much more modest package.

A consultant might be commanding a rate of several hundred euros a day, and a volunteer just making ends meet. These four people might be working on the same issues in a partnership between organisations.

The size of your pay cheque will depend on the organisation you are working for, your role and the type of contract (local recruit/international/consultant). *Local staff* can still be international, but recruited on a local contract without all the benefits that go with being *international*.

> As a Junior Professional Officer (JPO) working for the UN, I have a salary that is much higher than if I had an equivalent role back home in Europe. I am able to send about 40 per cent of my salary back home to pay off student debts and might even be able to put down a mortgage on a house in a few years. I even get a housing allowance. My salary now makes the years of struggle and poverty getting into the sector worthwhile and I am in a junior position. Those higher up the scale have very large salaries.
>
> *Anonymous*

As a general rule, the larger the bureaucracy, the higher the salary with internationally recruited staff (for UN, World Bank, Bilateral Government Donors, and Development Consultancy Companies) commanding the largest pay cheques. NGO salaries are less competitive as many fund their work through public funds or charitable donations and need to demonstrate how much is being used directly on projects, with pressure to reduce administration and management costs.

> Don't be fooled, there is a lot of money floating around within the development sector. The thing you need to discover is how to access it!
>
> *Anonymous*

But if money is your main motivation, you are better off working in the private sector for a company that offers large annual bonuses. You will never become rich or famous working in development or humanitarian assistance.

> When I am at home I buy a shirt for $5 in a charity shop to support their international operations. My friend, working for this organisation and based in Africa, receives a housing allowance that requires 400 of these shirts to be sold per month to cover personal housing costs.
>
> *Anonymous*

Risk, uncertainty and change

> Whenever I give a talk to people interested in working in humanitarian relief, my intention is to scare 80 per cent of the audience off the idea of a career in the sector. It is a dangerous and tough area to be working in: there is one death every three days and there is a high risk of kidnapping making it the fifth most dangerous sector (just behind things such as deep-sea fishing and logging). People working in the sector also have a very high risk of divorce.
>
> *Martin McCann, Chief Executive, RedR UK*

The humanitarian relief sector is not a field where you can get comfortable with the security and stability of your job or environment. Here you will often be working in insecure and dangerous environments, with tight security restrictions imposed on you, and evacuated at a moment's notice if the security worsens.

This applies to a lesser extent to development work although trends in development are constantly changing. One moment HIV and AIDS and gender were all the rage, next the demand is for business development, value chains and impacts evaluations. How do you make sure your skills stay updated and relevant? Here you will need to stay abreast of the latest development and adopt a lifelong-learning philosophy, not being scared to take on new challenges when they present themselves.

Development funding is equally fickle. Few projects are funded beyond five years, most work contracts will only be for a year or two, renewable. This is even more acute in humanitarian emergency work, where contracts may be as short as a few months at a time.

> When I do interviews to recruit my team I look for adaptability indicators. As a general rule those who have stayed in the same job in the same country for 12 years tend to be less adaptable than those who have taken on a new challenge in a new company every two to three years.
>
> *Silla (Hong Kong)*

Short-term contracts and general mobility of people within the sector do have their plusses though. It does ensure that you don't stagnate and that you become very open to change. This can make for a very exciting life, which might suit you when you are young, but what about when you are older, with a family? Will they be able to adapt with you, join you?

> When I return home and see my friends from university, I can genuinely say that, of everyone, I think I enjoy my job the most. My friends are all earning high salaries in 'the city' but can't stop complaining about commuting times, long working hours, London life in general and how they want to escape it. I like where I live and get genuine job satisfaction, if not their high salaries.
>
> *Hannah (UK)*

So, all things considered, is this a career for you?

Now that you have an insight into some of the highs and lows of this type of work, is it for you? Or would you be better off working in a different area?

You might get a closer look at indigenous communities by being a research anthropologist than a development worker; find a greater thrill out of being an adventure sports instructor than working in humanitarian assistance; travel – and get paid – more as an executive of a multinational company; and find greater job satisfaction helping the poor and disadvantaged in your own community or city.

PLATE 2 In much of Africa cooking is done using very basic technology, which has high environmental and health impacts. By training local entrepreneurs to use efficient cookstoves, Metamorfose can roll out these technologies and help others to understand the environmental and health benefits of improved cooking technologies. Metamorfose's focus is to develop local expertise so that this sector will become sustainable in the long term.

Source: Metamorfose/Line Loen

3

WHO COULD YOU WORK FOR?

When people think of working in development, non-governmental organisations (NGOs) are often the first organisations that come to mind. But the field of development and humanitarian assistance is broad, encompassing governments and bilateral aid agencies, multilateral agencies, development banks, NGOs – of all shapes and sizes – consultancy firms, universities, think tanks, research and knowledge platforms, social and private businesses, among others. When it comes to employers, it is wise to think as broadly as possible to enhance your options and find the most suitable match.

The first section of this chapter gives you an overview of the different categories of organisations working in development or humanitarian assistance, and the second part of the chapter helps you to create an organisational shortlist, allowing you to target organisations with a good match and where you can excel.

Types of organisations

Bilateral organisations (government aid agencies)

Examples include:

- Abu Dhabi Fund for Development
- Brazil – Agência Brasileira de Cooperação
- Chile – Agencia de Cooperacion Internacional de Chile (AGCI)
- Denmark – Danish International Development Agency (DANIDA)
- Netherlands – Embassy of the Kingdom of the Netherlands (EKN)
- France – Agence Française de Développement (AFD)
- Ireland – IrishAID
- Japan – Japanese International Cooperation Agency (JICA)

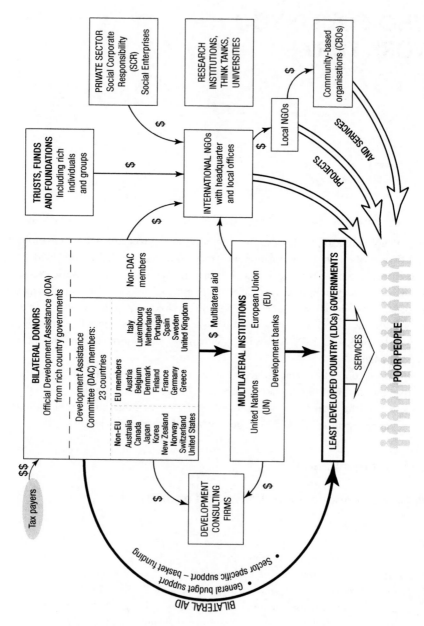

FIGURE 3.1 Aid architecture diagram.

- South Korea – Korea International Cooperation Agency (KOICA)
- Luxembourg – Luxembourg Development (LuxDev)
- Saudi-Arabia – The Saudi Fund for Development
- Spain – Agencia Española de Cooperacion Internacional y Desarrollo (AECID)
- Sweden – Swedish International Development Agency (SIDA)
- United Kingdom – Department for International Development (DFID) and UK Aid
- United States – USAID and the Millennium Challenge Cooperation (MCC).

Bilateral donors are the key players when it comes to funding development and humanitarian assistance. The Development Assistance Committee (DAC) is made up of 23 of the world's richest countries (see Figure 3.1) who, in 2010, provided around 90 per cent of the total aid disbursement, amounting to $129 billion.[1] Non DAC members also provide funding, with countries such as Turkey, for example, becoming an increasingly important player. Their funding is termed Official Development Assistance (ODA).

Administration of ODA varies: in some countries it is managed by the ministry of foreign affairs, in others by a specific agency focused on development that may be part of government or independent, or a combination of all three. The United States government for example, one of the largest providers of ODA, has more than a dozen agencies handling US development funding: USAID manages the bulk, disbursing around 56 per cent of the total US ODA (2009) with the MCC being the second largest. The Dutch government provides ODA through their ministry of foreign affairs. In the recent financial climate characterised by budget cuts, countries such as Canada and Australia are shifting administration of ODA from specific development agencies to their ministries of foreign affairs.

Individual countries' development assistance programmes differ greatly in terms of their development and country priorities according to historical ties (see Chapter 4 for further details). Despite efforts to coordinate, each donor makes its sovereign decisions regarding the objectives, priorities and incentive framework of its aid programme.[2] ODA is channelled through:

- multilateral organisations and multi-donor funds (such as the UN and the Global Fund or EU);
- direct support to developing country governments, through either general budget support or sector-specific budget support or project funding;
- NGOs or consultancy firms to deliver projects, including partnerships with the private sector (termed Public Private Partnership (PPP));
- loans.

In conflict or post-conflict countries, where governance is week, bilateral aid is usually channelled through nongovernmental organisations to ensure that it gets to the people and to avoid corruption. In more stable countries, budget support is more common.

Working for a bilateral donor you will help to manage that country's support and resources effectively and accountably as well as engage in policy development and support partner countries to achieve their development objectives. Roles will vary from administrative to advisory, operational or policy. In a recipient country receiving a general budget or sector-specific support, you might be advising and overseeing the correct utilisation of the funds. You could also be reviewing project proposals from NGOs and managing their contracts, or identifying projects to fund and outsourcing these to development consultancy firms. Coordinating with other donors, ensuring complementarity of funding rather than duplication is also important.

One advantage of working for a donor such as a bilateral organisation is that you don't have to chase funding and will have a larger influence over where the money is spent. As budgets are large, the level of influence can be greater. They can also offer more of a structured career path, professional development and job security but entry is often highly competitive. In most cases you will be a civil servant, although short-term consultants or technical assistants may also be recruited to input on specific programmes. Recruitment policies vary, and entry may involve competing in general exams or recruitment rounds or, for more technical positions, responding directly to a job vacancy. Most run internship programmes or training programmes. In many cases you will have to be a national of that country (or region, as is the case in the EU) in order to be eligible to work there.

> After six years of working in NGOs I decided I wanted to have more say and influence over where the money was going. I felt I could make more of a difference if I worked for a donor. I did a PhD to expand my knowledge which helped me to be recruited as an advisor in a multilateral agency. Now I try to make sure that poor people really do benefit from the funding and have a say in how the money is spent. The way I survive in a big bureaucracy is seeing the delivery of the programmes, and meeting the people who benefit.
>
> *Anonymous*

> I worked for Japan Bank for International Cooperation on infrastructure finance. The bank facilitates this finance scheme, providing low- and middle-income countries with funds to jointly build the infrastructure with private companies. I was involved in research and organising conferences, which gave me the chance to meet many senior people from the WB, UN, government and other development banks. However, I always felt unsure of the impact we were having, and could not see the immediate outcomes from the conference and research papers. I wanted to work on the ground and at a smaller scale, to be responsible for what I was doing and see the result of what I had done. I now work for a private firm in East Africa, providing markets and jobs to local people.
>
> *Momoko (Japan)*

Multilateral organisations and development banks

THE UN AND ITS AGENCIES

The UN was established in 1945 and has 192 member states, so its operations are understandably vast. It is made up of around 14,000 employees around the world. This does not include the operations of the World Bank and the International Monetary Fund (IMF) which are specialised UN agencies. The World Bank alone employs more than 10,000 people from over 160 countries – two-thirds at its headquarters in Washington, DC. A selection of UN agencies focusing on development and humanitarian issues includes:

* Food and Agriculture Organization (FAO)
* International Fund for Agricultural Development (IFAD)
* International Labour Organization (ILO)
* International Monetary Fund (IMF)
* International Organisation for Migration (IOM)
* United National Development Programme (UNDP)
* United Nations Educational, Scientific and Cultural Organization (UNESCO)
* United Nations Environment Programme (UNEP)
* United Nations Children's Fund (UNICEF)
* United Nations Industrial Development Organization (UNIDO)
* United Nations High Commissioner for Refugees (UNHCR)
* United Nations Office for Project Services (UNOPS)
* United Nations Population Fund (UNFPA)
* United Nations Office against Drugs and Crime (UNODC)
* World Bank Group
    – International Bank for Reconstruction and Development (IBRD)
    – International Finance Corporation (IFC)
    – International Development Association (IDA)
* World Food Programme (WFP)
* World Health Organization (WHO)
* World Tourism Organization (UNWTO)
* United Nations Entity for Gender Equality and the Empowerment of Women (UN Women)
* Joint United Nations Programme on HIV and AIDS (UNAIDS).

You can find more details on the UN system of organisations at www.unsystem.org/ and more about careers with the UN at https://careers.un.org. Each agency has its own website where vacancies are advertised and internship details given, as well as the employment process and the type of people they recruit.

Multilateral organisations are formed between three or more nations to work on issues that relate to all of the countries in the organisation. They are funded by the governments of their members states and therefore often have quotas on recruitment from different member states. This means that they provide very diverse workplaces. More than 200 multilateral agencies – such as the United Nations, the World Bank and the global funds – receive about one third of total ODA.[3] Reconstruction and development banks, under the World Bank, include African Development Bank (AfDB), Asian Development Bank (ADB) and Inter-American Development Bank.

The work within multilateral organisations is extremely broad. You might be involved in directly implementing large-scale projects, providing grants or loans to partner governments to execute projects themselves, policy development, coordinating humanitarian relief or managing the overall coordination of donor funding. While working within a complex bureaucracy, many multilateral organ-isations offer competitive compensation and ample advancement opportunities, so are attractive career paths for those at the entry, mid and senior levels. While full-time jobs are increasingly rare, part-time positions and short-term assignments are more common, so with the right connections you can get in through the back door. Skills and experience are important, and education is highly valued. Languages are important in many of the multilateral organisations as you will be working across countries and teams. Contrary to perception, the development banks do not just recruit economists but also people from many other areas of specialisation including of course economics, finance, human resource development (public health, education, nutrition, population), social sciences (anthropology, sociology), agriculture, environment, private sector development, as well as other related fields. The UN official working languages are Arabic, Chinese, English, French, Russian and Spanish and knowledge of at least two of these is vital. The UN has different categories of staff, each with its own requirements:

- Professional and higher categories
- General Service and related categories
- National Professional Officers
- Field Service
- Senior Appointments.

Many UN agencies have a Junior Professional Officer (JPO) scheme that is administered in some cases by the www.jposc.org/of the UNDP. To be eligible you have to be under 32 years of age and be a national of a country that sponsors JPO programmes at that time.

> When I heard that in 2005 Singapore was offering the United Nations National Competitive Recruitment Exam (NCRE), open annually to young professionals who are nationals of member states who need additional representation among staff in the United Nations Secretariat, I decided to go for it. After passing

the exam and the subsequent interview, I was offered a job in the Middle East. I worked in human resources previously and this tied in closely with my position within the UN. After two years in the post I then moved back to East Asia, to take up my second posting. The opportunity to live and work in different places is very appealing and my job challenges me and gives me the opportunity to learn a lot.

Anonymous

Non-governmental organisations (NGOs)

BRAC
The Wikimedia Foundation
Acumen Fund
Danish Refugee Council
Partners in Health
Ceres
CARE International
Médecins Sans Frontières
Cure Violence
Mercy Corps
Apopo
Root Capital
Handicap International
IRC
Barefoot College
Landesa
Ashoka
One Acre Fund
Clinton Health Access Initiative
Heifer International[4]

Over the past decades, NGOs have become major players in the field of international development, helping to shape the lives of millions of people around the world. They are responsible for channelling a significant amount of overseas development aid, are becoming an increasing global influence and are a dynamic, innovative and inspiring sector.

The term NGO is broad, and can be applied to any non-profit organisation that is independent of government. Although the NGO sector has become increasingly professional over the last two decades, the principles of altruism (charitable donations) and voluntarism remain key defining characteristics. NGOs come in many shapes, sizes and remits and include:

- *community-based* organisations (CBOs) – which serve a specific population in a narrow geographic area and are generally membership organisations made up of a group of individuals who have joined together to further their own interests (e.g. women's groups, credit circles, youth clubs, cooperatives and farmer associations); these may also be called grassroots organisations;

- *national* NGOs – which operate in individual countries; and
- *international* NGOs (INGOs) – which are typically headquartered in developed countries and carry out operations in more than one developing country. They usually engage with local NGOs and CBOs as partners in their work.

NGOs need to raise funds in order to implement their activities, so fundraising is a big part of their work. This does raise a number of criticisms, among them that NGOs are pandering to donor interests and may over-promise short-term successes without being able to deliver results in the long term. On the plus side, NGOs are seen as cost-effective (achieving a lot with a small budget when compared to multilateral or consultancy organisations), with strong grassroots links and field-based development expertise. They tend to engage in long-term commitments that have an emphasis on sustainability and have the freedom to innovate and adapt with a focus on participatory methodologies and tools. NGOs are also increasingly sharpening their business ethos, tightening their operations and management structures, enhancing their accountability and adopting some philosophies once considered a corporate mainstay, such as performance-based pay and cost–benefit analyses. Private sector experience can be a definite asset for those wanting to enter the NGO sector.

NGOs may be *operational*, with the primary purpose of designing and implementing development or humanitarian assistance-related projects, or *advocacy* NGOs – who work to defend or promote a specific cause and who seek to influence the policies and practices. A growing number of NGOs engage in both operational and advocacy activities.

NGOs vary in size dramatically, and are growing with the merger trend between organisations. But impact doesn't necessarily grow with size and smaller organisations also have a very important role to play in development. Working for a small NGO you will probably get involved in a whole range of areas and have more opportunities to engage in high-level activities at a relatively junior level. On the downside, smaller NGOs may have limited financial and management expertise. While you may adopt a high level of responsibility early on, you may experience some professional isolation, so support networks or mentoring outside the organisation are important.

> The advantage of working for a small NGO is that you get involved in everything. You have a very steep learning curve and you have the possibility to represent the organisation in a large number of forums.
>
> *Sophie (UK)*

NGOs hire in a variety of ways, and vacancies are often created as a response to new programme funding being secured, with increased competition among NGOs and drive for accountability and results. NGOs are innovating and most offer a dynamic working environment, with little bureaucracy and opportunities to engage closely with beneficiaries, designing, implementing and monitoring programmes. On the downside, due to limited capacity and endless opportunities, staff in many NGOs tend to be overworked (so welcome the support of interns and volunteers). Salaries within the NGO sector seem to be relatively low and contracts tend to be for a fixed term, according to project duration.

IN FOCUS: GOAL

GOAL is a Dublin-based international aid organisation that is currently operating in 13 countries across the developing world, the latest of these being Syria. The organisation's headquarters are at Dun Laoghaire, Co Dublin, and it maintains offices in London and New York.

GOAL delivers a range of programmes in the developing world, including emergency response, health care, water and sanitation, education and nutrition.

Of the almost 3,000 staff that GOAL employs, the vast majority are local to the country in which they are located. Besides implementing its own programmes, GOAL often works with partner organisations and missionary groups who have similar objectives. The organisation recruits according to its specific requirements at any given time, but is always keen to add accountants, nutritionists, logisticians and engineers to its ranks. At head office in particular, volunteers, interns and secondees from companies in the private sector also play a role within GOAL.

Trusts, funds and foundations

Aga Khan Development Network
African Capacity Building Foundation
Rajiv Gandhi Foundation
The Global Fund for Women
One Laptop Per Child Foundation and Association
Ford Foundation
Draper Richards Kaplan Foundation
Near East Foundation
Google.org Foundation
The Ashmore Foundation
Clinton Foundation
United Nations Foundation

Elton John AIDS Foundation
Damien Foundation
Children's Investment Fund Foundation
Bill and Melinda Gates Foundation
Touch Foundation
Edna McConnell Clark Foundation
National Lottery Fund

These organisations, philanthropic in nature, provide funding through grants to implementing agencies. There are two types: private foundations – which derive their money from a family, an individual or a corporation – and grant-making public charities which collect money from various sources to distribute.

Some foundations are more hands-on and identify their priority areas, what they would like to fund (be it education projects, HIV and AIDS awareness, emergency response), openly advertising funding rounds or calls for proposals. Once awards have been made they may actively support their grantees helping to build their capacity and undertake evaluations, and sometimes even implementing some of the work themselves. Here their work may be grant administration, supporting grantees, visiting the work, and reporting back to shareholders or trustees). Other trusts might be less open, supporting projects that come recommended by their trustees for example, and less involved in the implementation of the projects.

Size varies enormously, from small family trusts with a part-time employee, to very large operations, such as the UK National Lottery Fund or the Bill and Melinda Gates Foundation. NGOs are increasingly strengthening relations with large foundations as they are often flush with cash and offer more flexible funding than government funding. Corporate social responsibility (CSR) has also mushroomed in recent years with private firms establishing a charitable arm to either fund or implement projects. Some might partner with specific charities to support their work, while others might be more hands-on, funding the work of a partner NGO or civil society organisation (CSO). Some companies will employ people to run the CSR projects from within their existing staff, who may not have a background in development but know the company well, but increasingly organisations are looking for specialists in the area. Many are also increasingly involved in implementing projects with an increasing overlap with NGOs.

> I worked for my company's private foundation, started after the trustees agreed to donate a percentage of the profits to charitable causes. We like to fund small NGOs working in the field of education. My job involves identifying potential organisations and inviting them to submit an application, which I then present to the board for approval. I then arrange their contracts and oversee their reporting and spending. Increasingly I am spending more time in the field, visiting the work we are funding, and trying to quantify its value to our trustees.
>
> Anonymous

Development consultancies

Abt Associates
Creative Associates Inc.
Crown Agents
DAI
Deloitte Consulting
International Resource Group
John Snow Inc.
Landell Mills
Louise Breger Group
Tetra Tech ARD
Universal Research Co.
Oxford Policy Management

Development consultancies are in the business of selling their expertise to governments, companies and NGOs. They usually win work contracts through a competitive tendering process. They are not usually involved in the project identification, but will be involved in the design and implementation. A donor (such as USAID, the World Bank or the European Union) will identify a project they want to fund. This might be strengthening strategic purchasing in Kazakhstan. A tender is released. Interested consultancy firms prepare and submit their proposals. The company with the winning bid will be contracted to implement the programme. They may already have in-house expertise to do so, or need to contract an external firm, or team of private consultants to work on and deliver that project.

Unlike NGOs, they are driven by profit (although some may have a not-for-profit structure or channel this to other projects) and have a business structure. Clients are not always development donors, but they may also work for private firms or be contracted directly by the governments of developing countries.

Consultancy firms recruit in two ways. One is for staff members – both manager and specialist positions – and occasionally there will be graduate entry roles too. The second is for short-term consultants with skills in specialist areas according to project needs (see Chapter 10 on becoming a consultant).

Consultancy firms often offer long-term career prospects, recruiting the best and investing in their training to enhance staff retention. You will also have the opportunity to work on high-level projects, do a lot of international travel and be managing large budgets. On the downside, you may be one step removed from the final beneficiaries, and may not be able to influence project design, which may result in you implementing a programme you don't really believe in, but it is what the client has requested. You will also find that a significant amount of your time is spent preparing complex bids and tenders.

Launching a career in a development consultancy can be a good way to learn the essentials of programme management and donor contract compliance, to later make the transition to NGO management or as a consultant. Many offer internships and graduate training schemes or entry level programmes.

DID YOU KNOW?

USAID, the world's largest bilateral donor, awards millions of dollars each year to consultancies and other implementers. USAID gives out the largest sums of money in multiyear arrangements called indefinite quantity contracts (IQC), which are typically won by well-established US development consultancies. Those with multiple areas of expertise can often win several IQCs. Companies who are awarded IQCs can further subcontract to other businesses.

One of the largest private sector companies working with IQCs in the US is Chemonics, founded in 1975 and based in Washington DC. Entirely employee-owned and ISO-9001 certified, they work with the US Agency for International Development, other bilateral donors and private sector partners to design and implement development projects in many of the world's developing countries. Committed to their mission of helping people live healthier, more productive and more independent lives, they have worked in more than 140 countries in the areas of financial services, private sector development, health, environmental management, gender, conflict and disaster management, democracy and governance, and agriculture.

I work for a private development consultancy firm based in Madrid. Our projects are mostly funded by the EU after being awarded the services contract through international bidding. My organisation specialises in Social and Economic Development and we primarily work in North Africa, Latin America and West Africa.

I work on a 4 million euros project with Algeria, providing support to small and medium enterprises (SMEs). We recruit consultants to help deliver the work so a large part of my role is identifying the right experts as well as undertaking visits to the partner countries to monitor work, organise the logistics, present the team and discuss expectation and achievements with the client (in this case the Ministry of Industry and Enterprises).

My favourite part of the job is dealing with experts from different backgrounds, speaking three languages every day and maintaining diplomatic relations with all parties involved. My least favourite aspect of the job is the financial side: negotiating fee rates with an expert who wants 600 euros a day is always tough. One of my favourite aspects of the job is travelling to beneficiary countries.

Estelle (France)

Think tanks, institutes and universities

Think tanks are public policy research institutes that seek to play a key role in making and influencing global policy through advocacy. They aim to improve their respective spectrums, as well as being sources of new ideas and research. Many thinks tanks actually are 'think and do tanks' in that they also take on specific consultancies or sometimes even implement projects. Universities are moving this way too, combining academic research with consultancy work, which gives them a more practical orientation. Knowledge platforms and networks that bring different key players together are also potential employers.

The majority of think tanks are quite small and therefore there are not many vacancies at any one time. Employers generally look for detailed knowledge of research techniques, so a taught Master's degree course in social research methods or obtaining a research degree in your particular field are very useful. Some examples of think tanks and research institutes in the UK include:

- Overseas Development Institute (ODI)
- Legatum Institute
- Institute of Development Studies (IDS)
- Chatham House: Royal Institute for International Affairs
- International Institute for Environment and Development.

Some institutes such as the Institute of Development Studies (IDS) in Sussex (UK) or the Centre for Development Innovation (CDI) at Wageningen (The Netherlands) also offer academic teaching, and staff combine this with their own research projects and consultancy work.

Private sector

While the private sector was not considered a typical employer for those who wanted to work in development a few years ago, it is increasingly being recognised that the private sector can significantly contribute to poverty reduction. Market approaches to development combine the power of development and the social competence of NGOs and governments and improve the delivery systems for aid.[5] This can include a variety of roles and employers, ranging from CSR departments with for-profit companies, to companies with a social mission (social enterprises), to private companies who work in developing countries. The tea buyer in India or Uganda for example can play an important role, providing a market, influencing prices and working conditions, and able to push for fair trade practices. The telecommunications or banking manager expanding services and products to rural communities also has an important role to play, as do entrepreneurs investing in the country: creating jobs and disseminating new technologies plays a vital role in development. The number of companies that work in or with lower income countries are many – and growing in number – and there are some exciting opportunities within them. Being a socially minded person within these companies can be a huge asset both to your clients and the companies. Social enterprises – companies with a social remit – are also interesting employers, as their overlap to development is a large area, with market-based approaches worth exploring. Trade has much greater potential in the long run to influence development than donor money.

After my master's in International Development and searching for jobs in the NGO sector without success I decided to broaden my search and came across jobs in the CSR sector. While competition was still fierce (many of the fellow applicants had PhDs, etc.) I had more luck, and eventually came out to Rwanda to manage the CSR projects of a small family coffee company. We supported the development of projects around the communities where the farmers received small grants. I managed and oversaw the grants, travelling out to visit the partners regularly.

Anna (USA)

Military and security forces

While not involved directly in development projects, military and security forces are among the first responders to humanitarian crises and also play a role in peacekeeping missions. If this is your area of interest, this may be an employer worth exploring.

CAREER CLINIC! HOW EASY IS IT TO MOVE BETWEEN TYPES OF ORGANISATIONS?

Many people start off with NGOs and then 'graduate' on to the UN or donors. There is a lot of merit to getting some initial hands-on NGO experience, gaining understanding at the *front line*. Some do the reverse, starting off with donors and transitioning later on to NGOs. This second route can be easier, as NGOs value donor, private sector, research and consultancy experience highly, if combined with knowledge on the ground. A long career in smaller NGOs can make the transition to other organisations harder, partly as other employers may – rightly or wrongly – view NGOs as lacking rigour and challenges within larger bureaucracies.

Inevitably it is always easier to go for other jobs in similar types of organisations – partly due to the understanding, experience and contacts you will have built up about these organisations. But if, mid career, you are set on making a transition, by cultivating the skill set, knowledge and understanding of opportunities (plus don't forget about nurturing relevant contacts) within your target organisations, the transition is possible.

Developing an organisational shortlist

Given the huge range of potential employers out there, it is helpful to develop a shortlist of organisations – or types of organisations – you would like to work for. This will enable you to understand the profile of the people they recruit, and thus better plan your entry. The benefits of working for the right type of organisation – where there is a mutual fit – cannot be overemphasised. Your job satisfaction, effectiveness in your role, motivation and your career progression will ultimately depend on this, and organisational culture varies greatly. Are you a brown bread and sandals person who likes to work closely with local communities and get down to action? Or do you prefer the high-level corporate crossover, attending conventions that bring together the major world players to debate and create policy resolutions?

When developing an organisational shortlist, it is important to base it – as much as possible – on reality: facts and experience. It is not a task that can be done over-night and you may have to constantly be reassessing what is right for you as you change jobs and move forward. It is easy to get drawn in by the superficial image of an organisation, aiming your goals in a direction that won't necessarily suit you. The UN is a classic example, a visible organisation with clout, good pay scales, and possibilities. Some people may thrive in this environment, while others may feel frustrated and sink within the bureaucracy.

At an early stage, reading about the experiences of people who work within the organisation on the organisation's website, is a good start. Doing some work experience, or an internship, is a better way of seeing an organisation from the inside. And as you move throughout your career you will have the chance to continually evaluate and assess different organisations, whether you are happy there, or choose to move to a different type. Here are some important considerations that will help you to prioritise organisations. Some inevitably draw on stereotypes, which do not always – but often – hold true.

A donor or an implementer?

Organisations can broadly be categorised as donor organisations versus implement-ing agencies. Donors provide the funding for development and humanitarian projects, be it through government(s), private foundations or CSR funds for example. Working for a donor you will be able to influence how and where the money goes. Donors decide what they want to fund – so there are some research and policy roles – and then award grants or contracts and oversee the implementation of the work, ensuring contractual obligations are met. Implementing agencies apply and compete for donor funding to deliver the projects. Here you are able to engage in programme design and delivery and are likely to work more closely with beneficiaries. These include INGOs, NGOs, CBOs, Consulting Firms and Think Tanks, for example.

Think about whether you are more interested in the big-picture work such as attending large conferences and analysing the effectiveness of development strategies and policy advice? Or do you prioritise working with, and understanding the local communities? In this case you are likely to be more suited to an NGO role, or perhaps in a practical action–research organisation.

Large or small organisation?

The experiences and opportunities will vary greatly according to the size of the organisation that you join. Smaller organisations have the great advantage of enabling you to work relatively independently, across a broad area of work (with fewer specialists, roles tend to be broad) assuming big responsibilities early on. This will enable you to learn on the job, and be influential from a relatively early stage. They tend also to be more flexible and responsive.

Conversely, larger organisations will likely offer more structure (with associated policies and procedures), professional training opportunities and ability to learn from a range of experienced people, even if your role is relatively restricted. Starting out, it may take longer to get engaged in interesting opportunities but there may be more avenues to progress to and have a structured career.

Think about whether you thrive in an environment where there is structure and boundaries to your work, or in a more creative and flexible environment that enables you to explore and take up opportunities.

> I worked at the HQ of a large INGO for two years before being field based. During those two years I made the most of all the training and development opportunities available in related topics such as standards in humanitarian response, protection and gender, as well as good field management in finance and human resources.
>
> *Juliette (UK)*

Organisational values/culture

Most organisations, especially NGOs are typically value based and as an employee it is important that you are sympathetic or directly share their values. Some may be faith-based organisations, others may have a specific standpoint or belief. For example, the Christian INGO, Tearfund, specifies that employees must be *Christ-centred*, whereas the Catholic Agency for Overseas Development (CAFOD) and IslamAid do not specify this. Some might be pro-abortion, while others may be strongly against this. Firmly believing in their remit will make you more passionate about their work.

Values can also determine how success is measured, which is important for your work. What are you working towards, and what gives value to your work? For example, in the private sector, profit is often the bottom line. For an NGO working with child sponsorship, number of children sponsored might be the indicator, or for another impact may be a priority.

Once you have developed your organisational shortlist, you can:

- check the profile of people on their staff and vacancy listings to help you plan your training and career path;
- arrange informational interviews with their HR departments;
- target these organisations whenever you are looking for volunteer opportunites, internships or entry level jobs.

My parents were diplomats so I was exposed to their work and as a result was very idealistic about internationalism and the UN when I was young. Because of my interest in conflict resolution I was drawn to the work of the UN Secretariat's Department of Special Political Affairs. I did a postgraduate degree in International Law to prepare me for work in this field. Whatever our plans, life has a way of taking its own course, and in the event I never got recruited into the UN Secretariat. In retrospect, this turn of events was probably for the better, as I would not have thrived in the culture and working environment there; nor perhaps would it have been the best place for a young professional to form modern and dynamic management practices.

I gave up a PhD at the age of 27, being impatient with the Ivory Tower, and, seeking to 'make a difference' in a world in need, I applied for a post as a JPO to work for UNDP, for which I was selected. This was a departure for me, as my initial interest was political not developmental. I spoke English, Spanish, French and Italian. I remember the interviewer was Mexican. They wanted to send me to Honduras but I said I wanted to be somewhere where poverty was deep and the needs were great. They sent me to Cameroon, a bilingual country. I thrived in this environment, and my career then took me to New York Headquarters, Ecuador, Armenia, the Philippines, Washington DC, Jordan and subsequently to Jerusalem in the occupied Palestinian territory, with The United Nations Relief and Works Agency for Palestine Refugees in the Near East (UNRWA). By any measure I progressed very rapidly in the career, and certainly by the standards of those days. At the age of 38 I was appointed the then youngest UNDP Resident Representative and had the privilege of opening the first UN office in Armenia. Nonetheless, although I closely identified with the UN and UNDP and its mission and aspirations – indeed I loved the UN deeply, and always will – I felt the gap between the official narrative and the quality of work on the ground was wider than I could live with. It was time to move on.

I went in search of more rigour in my work in the world of multilateral development banks (MDBs). I joined the Inter-American Development Bank at its Washington, DC Headquarters, the oldest regional development bank. Here, the average technical quality of the staff was relatively high. A PhD in Economics tends to be the norm, or at least the expectation. The atmosphere is serious and rigorous, as well as rather cautious. I was advisor to the president in strategic planning, and helped establish the office of internal evaluation, bolstering the ability of the bank to be accountable. But after a few years I missed the 'soul' and so chose to move again. In a career where I had played development diplomat in the UN system, and now development banker, my experience and outlook sorely lacked NGO experience. I wanted to see what that world was like. Personally, also, my wife and I (she also had had a career as a UNDP professional) wanted our young family to grow up in Europe. This time I joined the International Planned Parenthood Federation, one of the world's largest NGOs as Chef de Cabinet. I certainly found the passion I was looking for, in an institution with a 'values-driven' culture where social activists prevailed.

Eventually I joined The Vaccine Fund/GAVI Alliance as its first Chief Operating Officer (COO). Initially launched by Bill Gates with a visionary donation of $1.5 billion, we built it into a $2.6 billion operation with a board chaired by Nelson Mandela. I now have my own consultancy firm based in the French Alps, working mostly in the area of organisational development and programme evaluations, helping governments and institutions to become more effective. You will only excel in your career, and, far more importantly, realise yourself as a person, if you find your niche. Your 'niche' is not just this job or that. It's also a function of the human relationships you nurture, you invest in with sincerity and passion, with supervisors, colleagues and those you are meant to serve and support. But your niche will change over time and experience, because you yourself change. Don't be scared to go in search of what is missing. Especially, don't be scared.

Fabian (Canada/UK/world citizen)

The worksheet in Figure 3.2 will help you to develop an organisational shortlist. Once you have done this you can:

- check the profile of people they recruit to help you plan your training and career path;
- arrange informational interviews with their HR departments;
- target these organisations whenever you are looking for internships.

CAREER CLINIC! HOW TO CONDUCT AN INFORMATIONAL INTERVIEW

A great way to explore different career options is to meet with people whose job you think you might like to do. This strategy is useful throughout your career, not just at the start. It will help you to gain knowledge about the profession, skills requirement and work in the field as well as get your name out there. The easiest way to get an informational interview is through personal referrals or contacts; or alternatively through alumni or professional networks. You could approach HR departments directly although they are likely to be very busy and have little or no time for you. Contact people within the department where you might be interested in working by targeting individual people by email. If they don't answer don't persist, although it is acceptable to send one follow-up email. Some questions you might like to ask during the interview include:

- What jobs and experiences have led you to your present position? What do you do? What are the duties/functions/responsibilities of your job?
- What decisions do you make/which problems do you deal with?
- What future employment trends do you see in the sector?

Prepare a good list of questions, be aware that they won't have much time for you, and above all do not ask for a job but you can explore opprtunities.

WORKSHEET: Chapter 3. Developing an Organisational Shortlist

1) Which organisational values are most important to you? List what you are looking for in an employer below.

_____   _____   _____

_____   _____   _____

_____   _____   _____

2) Which types of organisations appeal the most and why? Choose 2 from the below and consider why these organisations appeal to you:

☐ Bilateral organisations   ☐ Multilateral organisations

☐ NGOs   ☐ Trusts, funds or foundations

☐ Development Consultancies   ☐ Think tanks, institutes or universities

☐ Private Sector   ☐ Military/Security forces

Organisation type:.......................................   Organisation type:.......................................

....

Why:.......................................   Why:..

....

....

....

3) Research organisations within your chosen categories. List four organisations of interest within your chosen fields.

1) a _____   b _____   c _____   d _____

2) a _____   b _____   c _____   d _____

4) Try and understand who each of these organisations recruit, what skills, competencies and experience they are looking for in their candidates. N.B. Looking at job vacancy adverts and the CVs (if available) of existing staff should help you with this.

....

....

....

FIGURE 3.2 Worksheet: developing an organisational shortlist.

Notes

1 OECD data 2010.
2 Identification and Monitoring of Potentially Under-aided Countries, OECD REPORT; www.oecd.org/dac/aid-architecture/Identification%20and%20Monitoring%20of%20 Potentially%20Under-Aided%20Countries.pdf (accessed 30 January 2015).
3 www.oecd.org/dac/aid-architecture/multilateralaid.htm (accessed 30 January 2015).
4 Top 20 NGOs in the Top 100 NGOs ranking 2013 conducted by the *Global Journal* http://theglobaljournal.net/group/top-100-ngos/ (accessed 30 January 2015).
5 Urs Heierli (2008) *Market Approaches to Development,* Swiss Agency for Development and Cooperation (SDC), Employment and Income Division; Urs Heierli (MSD Consulting), Berne 1st Edition: printed in India. Hard copies are available at SDC, Employment and Income: e-i@deza.admin.ch Electronic copies: can be downloaded from: www. poverty.ch
This publication is supported by Employment and Income Division of Swiss Agency for Development and Co-operation – SDC, Freiburgstrasse 130, CH – 3003 Berne, Switzerland www.deza.admin.ch/Themes

4

WHERE COULD YOU WORK?

Opportunities for international travel, to live and work in a country other than your own are one of the attractions of international development and humanitarian work. You may be based either at an organisation's HQ or at a field office. You could be based in your own country or be working internationally. There are also roving positions, working across many country offices, contributing your expertise as needed.

The purpose of this chapter is to give you an understanding of where the jobs will be concentrated, and what it can be like to live and work in different countries and regions. The final section focuses on some of the peculiarities of working in a humanitarian emergency.

> When I started out in development, few people wanted to work in deprived locations, as developing countries were perceived to be, so there was a high salary and many benefits. These days, the pendulum has swung the other way, with international positions being some of the most competitive. This is especially true now that the emphasis is on local recruits, even for senior positions where possible, creating fewer jobs for international staff.
>
> *Geoff (Australia)*

Where are the jobs?

Jobs follow the funding, which will be concentrated in:

- donor countries who contribute the largest amounts of ODA;
- recipient countries who receive the largest amounts of ODA.

While donor countries channel their ODA to recipient countries directly, they also fund organisations (e.g. NGOs, development consultancies, research institutes, think tanks) with headquarters in the donor country. The countries with the largest disbursements of ODA are also the ones where you are most likely to find the greatest number of jobs (and development organisations). The top seven countries with the largest disbursement of ODA are listed in Table 4.1.

The list in Table 4.1 is closely followed by Canada, Sweden, Australia, Denmark, Germany, Belgium, Switzerland, Finland, Ireland, Italy, Korea, Luxembourg, New Zealand, Australia, Greece and Portugal. So jobs will be located in organisations headquartered in these countries to help manage funding and aid programmes with partner countries. In recent years many OECD countries have faced budget austerity measures that have resulted in aid cuts, freezing or reducing levels of ODA.[1]

Where is this funding channelled? Aid is primarily channelled to least developed countries (LDCs) and low income countries (LICs) and each donor makes

TABLE 4.1 Top seven bilateral donors by aid disbursement (2009 figures)[a]

Country	Total ODA	Recipient countries
United States of America (USA)	$14.7 billion (100% grants)	120 countries worldwide
Japan International Cooperation Agency	$9.16 billion (of which $1.66 billion is in grants and $ 7.50 billion is in loans)	72 countries worldwide
UK Department for International Development (DFID)	$6 billion (100% grants)	27 countries in Africa, Asia and the Middle East
Dutch Ministry of Foreign Affairs	$4.92 billion (100% grants)	Bangladesh, Benin, Burundi, Ethiopia, Ghana, Indonesia, Kenya, Mali, Mozambique, Palestinian territories, Rwanda, Sudan, Uganda and Yemen
Norwegian Ministry of Foreign Affairs	$3.08 billion	114 but 32 are 'important countries for Norwegian development cooperation'
Spanish Ministry of Foreign Affairs and Cooperation	Total ODA: $2.91 billion (of which $ 2.84 billion in grants and $71.3 million in loans)	50 countries but prioritises 23 in Asia, Africa, Latin America and the Middle East
Agence Française du Développement	Total ODA: $2.38 billion (of which $988.6 million in grants and $1.4 billion on loans)	65 countries worldwide

a Many countries started cutting ODA after 2009 due to the economic crisis. In 2009, for example, Spain's ODA represented the peak of 0.46% of gross national income (GNI) while in 2012 this reduced to 0.15% (a level last seen in 1989) (Source: http://donortracker.org/donor-profiles/spain).

its own decisions as to which countries are funding priorities, and the sectors within these it will emphasise. Aid architecture is a complex process, and despite efforts of the global development cooperation system to coordinate (the Accra Agenda for Action (AAA) in 2008 committed to 'improve allocation of resources across countries'), aid allocation practices remain to a large extent un-coordinated. Thus, some countries are *donor darlings* (development hotspots, countries with a large concentration of funding and some overlaps[2]) while others are *donor-orphans*[3] (under-aided LDCs and LICs).

The recipients of ODI can be divided into the following categories:

- LDCs
- other LICs
- lower-middle income countries and territories
- upper-middle income countries and territories.

Donors also create additional categories to plan their allocation of funds, including landlocked developing countries, small island developing states, fragile states (those that are mid conflict or have recently emerged from a conflict) and emerging economies such as or BRICS (acronym developed from Brazil, Russia, India, China and South Africa but now generally used to describe countries with rapidly emerging industries and increasing wealth).

DID YOU KNOW? LEAST DEVELOPED COUNTRIES

There are 49 countries currently classified as LDCs, which is a ranking of the world's most impoverished countries based on socio-economic analysis.[4] The criteria have been refined over the years to take into account new insights from research on economic development, updated information on structural impediments to development and improvements in the availability of internationally comparable data. Currently, the following criteria are used to classify countries as least developed:

- GNI per capita (average GNI of $905.12);
- Human Assets Index, which reflects the following dimensions of the state of human development including (a) Health and nutrition, measured by: (i) percentage of the population undernourished; and (ii) under-five child mortality rate; and (b) Education, measured by: (i) gross secondary school enrolment ratio; and (ii) adult literacy rate;
- Economic Vulnerability Index, a measure of risk posed to a country's development by exogenous shocks.

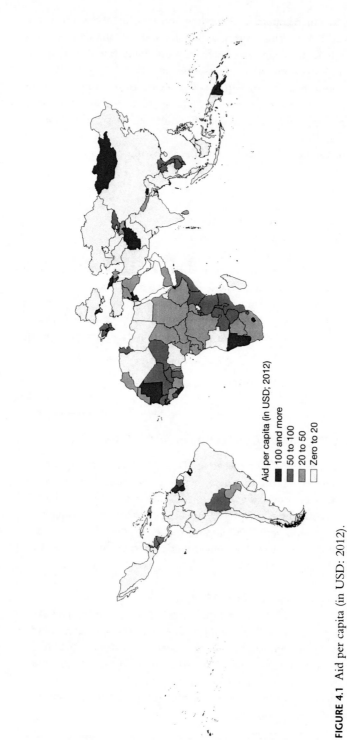

FIGURE 4.1 Aid per capita (in USD: 2012).

Source: Reproduced with permission from OECD (2012), *Identification and Monitoring of Potentially Under-aided Countries*, www.oecd.org/dac/aid-architecture/fragmentation-orphans.htm

TABLE 4.2 DAC list of ODA recipients: effective for reporting on 2012 and 2013 flows

Least developed countries	Other low income countries (per capita GNI ≤ $1,005 in 2010)	Lower-middle income countries and territories (per capita GNI $1,006–3,975 in 2010)	Upper-middle income countries and territories (per capita GNI $3,976–12,275 in 2010)
Afghanistan	Kenya	Armenia	Albania
Angola	Korea, Democratic	Belize	Algeria
Bangladesh	Republic of	Bolivia	★Anguilla
Benin	Kyrgyzstan	Cameroon	Antigua and
Bhutan	Republic	Cape Verde	Barbuda
Burkina Faso	Tajikistan	Congo, Republic	Argentina
Burundi	Zimbabwe	of the	Azerbaijan
Cambodia		Côte d'Ivoire	Belarus
Central African		Egypt	Bosnia and
Republic		El Salvador	Herzegovina
Chad		Fiji	Botswana
Comoros		Georgia	Brazil
Congo, Democratic		Ghana	Chile
Republic of the		Guatemala	China
Djibouti		Guyana	Colombia
Equatorial Guinea		Honduras	Cook Islands
Eritrea		India	Costa Rica
Ethiopia		Indonesia	Cuba
Gambia		Iraq	Dominica
Guinea		Kosovo	Dominican Republic
Guinea-Bissau		Marshall Islands	Ecuador
Haiti		Micronesia,	Former Yugoslav
Kiribati		Federated States	Republic of
Laos		Moldova	Macedonia
Lesotho		Mongolia	Gabon
Liberia		Morocco	Grenada
Madagascar		Nicaragua	Iran
Malawi		Nigeria	Jamaica
Mali		Pakistan	Jordan
Mauritania		Papua New Guinea	Kazakhstan
Mozambique		Paraguay	Lebanon
Myanmar		Philippines	Libya
Nepal		Sri Lanka	Malaysia
Niger		Swaziland	Maldives
Rwanda		Syria	Mauritius
Samoa		★Tokelau	Mexico
São Tomé and		Tonga	Montenegro
Príncipe		Turkmenistan	★Montserrat
Senegal		Ukraine	Namibia
Sierra Leone		Uzbekistan	Nauru
Solomon Islands		Vietnam	Niue
Somalia		West Bank and	Palau
South Sudan		Gaza Strip	Panama

continued . . .

TABLE 4.2 Continued

Least developed countries	Other low income countries (per capita GNI ≤ $1,005 in 2010)	Lower-middle income countries and territories (per capita GNI $1,006–3,975 in 2010)	Upper-middle income countries and territories (per capita GNI $3,976–12,275 in 2010)
Sudan			Peru
Tanzania			Serbia
Timor-Leste			Seychelles
Togo			South Africa
Tuvalu			★St Helena
Uganda			St Kitts-Nevis
Vanuatu			St Lucia
Yemen			St Vincent and
Zambia			Grenadines
			Suriname
			Thailand
			Tunisia
			Turkey
			Uruguay
			Venezuela
			★Wallis and
			Futuna

★ Territory.

Source: Reproduced with permission from OECD. OECD DAC List of ODA Recipients Effective for reporting on 2012 and 2013 flows www.oecd.org/dac/stats/DAC%20List%20used%20for%202012% 20and%202013%20flows.pdf

Aid allocation to recipient countries and within LDCs is unevenly distributed (Figure 4.1; Table 4.2). For example, Mozambique and Madagascar are similar both in terms of the size of their population, income per capita and poverty levels. In 2010, the population of both Madagascar and Mozambique was around 22 million with a GNI per capita of $430 (Madagascar) and $440 (Mozambique). The latest poverty estimates of the proportion of people living on less than $2 per day was 93 per cent for Madagascar and 82 per cent for Mozambique. Nevertheless, Mozambique received nearly four times more aid than Madagascar. Similarly, over the past five years, Tanzania has received nearly 60 per cent more aid than Kenya despite similar country characteristics.[5]

Geographical regions

This section looks at some of the principal regions you might be working in.

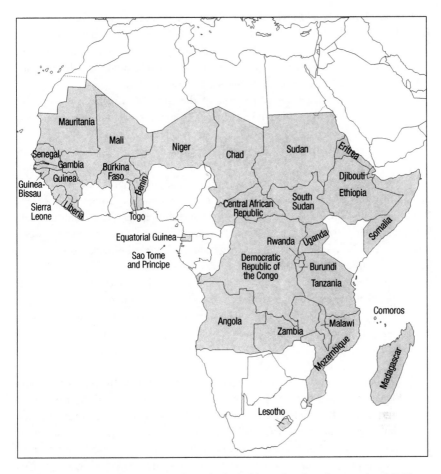

FIGURE 4.2 Sub-Saharan Africa map with shaded least developed countries (LDC).

Sub-Saharan Africa (SSA)

For those with little knowledge of the region, Africa is often synonymous with poverty, drought, famine, HIV, corruption and ethnic violence. Admittedly these are all issues, but the picture is not uniform. Despite the difficult global economic environment, growth in Africa has averaged at about has averaged at about 5 per cent per year (www.africaneconomicoutlook.org/statistics/table-2-real-gdp-growth-rates-2003-2013/) and some resource-rich or well-managed countries have had an average growth rate of 7 per cent or more.

In SSA all except two countries were under European colonial power in the nineteenth century 'scramble' for Africa. Only Liberia, an independent nation settled largely by African Americans, and Ethiopia, an Orthodox Christian nation formerly known in Europe as Abyssinia were untouched by colonial rule. SSA is the largest recipient of aid among OECD countries, receiving approximately *36 per cent of total global ODA.* Over the past four decades, aid to Africa has quadrupled from around US$11 billion to US$44 billion.[6] In 2008 the top five countries receiving the largest amount of ODA were Ethiopia (8 per cent of total), Sudan, Tanzania, Mozambique (each receiving 5 per cent of total) and Uganda and Democratic Republic of the Congo (each receiving 4 per cent of total).

ODA makes up a significant portion of the national income in many SSA countries. Countries most dependent on ODA are Liberia where it makes up 175 per cent of GDP, Burundi (40 per cent of GDP), Democratic Republic of the Congo (29 per cent), Sao Tomé and Príncipe (24 per cent) and Sierra Leone (24 per cent).

LDCs: 34 of the 49 LDCs are in Africa: these are Angola, Benin, Burkina Faso, Burundi, Central African Republic, Chad, Comoros, Congo (Democratic Republic), Djibouti, Equatorial Guinea, Eritrea, Ethiopia, Guinea, Guinea-Bissau, Lesotho, Liberia, Madagascar, Malawi, Mali, Mauritania, Mozambique, Niger, Rwanda, Sao Tomé and Príncipe, Senegal, Sierra Leone, Somalia, South Sudan, Sudan, Tanzania, The Gambia, Togo, Uganda and Zambia (see Figure 4.2).

While education and capacity is increasing on the continent, one of the largest problems is knowledge and know-how and consequently it is a hot spot for a lot of jobs for external experts. A large proportion of those who work in international development are likely, at some time or other, to step foot and work in Africa.

> I worked in eastern Democratic Republic of the Congo in the North and South Kivu provinces, implementing a gender-based violence (GBV) prevention programme. The challenges of working in DRC are many. First there are practical challenges. Four of the six sites I was managing had no or a very bad phone network. When teams are on the ground it is difficult to support them and get things done if you can't communicate with them quickly. Access is also difficult – in the rainy season some roads are inaccessible. Other locations have no roads, or they are dangerous to travel on, so the only way in is by helicopter. Then there are cultural challenges. Working on GBV, in a country where women are often viewed as having few or no rights (legally they have rights, practically they don't), requires long-term cultural change. For example, if a husband dies, the wife cannot inherit. It goes to the eldest son who may throw her out. There is also a lack of a functioning legal system. There is a law but corruption is rife. There are very few courts, which are far away, with survivors having to pay for their travel there. Often for rape there is an out of court settlement – at least that way the family gets something: a goat or a case of beer. Sometimes a woman is forced to marry the rapist. If you encourage the victim to go to the police, the perpetrator may be free the next day due to corruption, which puts the victim at a greater risk. The work I was doing was development, but in an area of high insecurity; the communities we work with might be displaced from one day to the next. Our staff also had to be frequently relocated.
>
> *Alessia (Italy)*

DID YOU KNOW . . .? SOME FACTS ABOUT AFRICA

Nigeria has the largest population in sub-Saharan Africa (154.7 million people) and accounts for 18 per cent of the continent's total population.

Rwanda has the highest number of women in national parliament with 56 per cent of total seats. Comoros has the lowest with 3 per cent.

Contraceptive use (any method) is highest in Mauritius at 76 per cent; lowest is Chad at 3 per cent during the period 2000–09. (MDG 6).

In Chad, 209 out of 1,000 children die before the age of one; in Seychelles the rate is 12 per 1,000. (MDG 4)

In Rwanda it took three days to start a business (one day in 2015) and in Guinea-Bissau 216 days (in 2010).

The Second Congo War, which began in 1998 and involved eight African nations, was the largest war in African history. An estimated 5.4 million people died as a result of the war and its aftermath, making it the deadliest worldwide conflict since the Second World War. The war officially ended in 2006, but hostilities still continue today.

One of the oldest universities in the world is in Timbuktu, Mali. By the twelfth century Timbuktu was home to three universities. Over 25,000 students attended one of the Timbuktu universities in the twelfth century.

Source: *50 Things You Didn't Know about Africa*, the World Bank http://go.worldbank.org/IH6NLHRGY0

FIND OUT MORE!

Richard Dowden (2009) *Africa: Altered States, Ordinary Miracles*, London: Portobello Books.
Wangari Maathai (2010) *The Challenge for Africa*, London: William Heinemann.
Martin Meredith (2005) *The State of Africa: A History of the Continent Since Independence*, London: The Free Press.

Asia

The Asian Tigers of Hong Kong, Singapore, South Korea and Taiwan sustained rapid economic growth rates (in excess of 7 per cent a year) and rapid industrial-isation between the early 1960s and 1990s. The forces driving this growth included new technology, globalisation and market-oriented reform. This rapid growth helped to lift hundreds of millions out of extreme poverty, but population growth remains a challenge and the region is home to two-thirds of the world's poor, with more than 800 million Asians still living on less than $1.25 a day and 1.7 billion surviving on less than $2 a day. Poverty reduction remains a daunting task and the question of how to bridge the gap between rich and poor remains a challenge. In addition, the visible side effects of environmental damage are alarming. The reliance on fossil fuels has degraded air quality and ecosystems, reduced the supply of clean water and created significant health hazards. The total amount of ODA was $36,773 million (2010 data) which represents *28 per cent of total ODA* disbursed to 36 of the 49 countries in the region.

LDCs: Eight countries in Asia fall within the least developed category and as a result get special attention (Afghanistan, Bangladesh, Bhutan, Cambodia, Lao People's Democratic Republic, Myanmar, Nepal and Timor-Leste) (see Figure 4.3).

The top four recipients of ODA in 2010 in Asia were Afghanistan ($6,426 million[7]), Pakistan ($3,013 million), Vietnam ($2,940 million) and India ($2,806 million).

However, on a per capita basis, Asia received the lowest amount, $10 per person compared to the average for all developing countries of $24 per person, and compared to Oceania, which receives the highest amount, $221 per person.

The pool of home-grown talent is broader and deeper in much of Asia (compared to SSA). This is good as those resources are more likely to stay in the community and to contribute their talent long after a project has ended. But it is bad news for international development job candidates from Europe or North America, because their skills are often less in demand than other regions such as Africa.

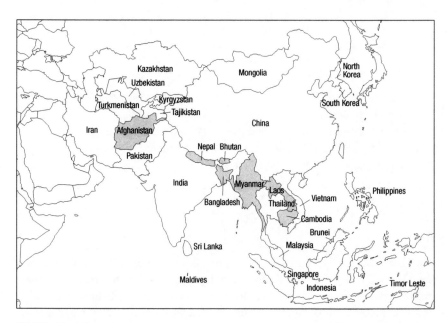

FIGURE 4.3 Asia map.

FIND OUT MORE!

Costas Lapavitsas (2012) *Beyond Market-Driven Development: Drawing on the Experience of Asia and Latin America,* Routledge Studies in Development Economics, London: Routledge.

Tom Miller (2012) *China's Urban Billion: The Story Behind the Biggest Migration in Human History,* London/New York: Zed Books.

Joe Studwell (2013) *How Asia Works: Success and Failure in the World's Most Dynamic Region,* London: Profile Books.

Middle East and North Africa (MENA region)

The MENA is a region of rich historical, cultural and religious heritage. With significant human and financial resources there is also a high level of infrastructure development. Extreme poverty in the region is low, with only 2.7 per cent of the MENA population living below 1.25 US$. High economic growth rates had a positive effect on the fight against poverty until the onset of the Arab Spring in January 2011. This in turn presented opportunities for political reform, and for greater political and economic freedom. One development focus is enabling women and youth to share development benefits. Like many low- and middle-income countries (LMICs), around two-thirds of MENA's population is under the age of 30. Youth need greater access to economic opportunities, quality education, recreation, and political participation. This lack of access, in combination with rising expectations brought about by education and the information revolution, creates frustration among youth and may even threaten the social fabric.[8]

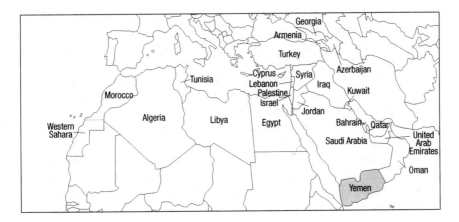

FIGURE 4.4 MENA region map.

There is only one LDC country in the region, Yemen, with most others being classed in the lower and middle income category (see Figure 4.4). Around *10 per cent of ODA* is channelled to this region.

> As a Western woman working in the Arab world I am often seen as a third gender. The rules that apply to men don't apply to me; neither do the rules of the local women. I am seen as a strange class of in-between.
>
> *Anonymous*

Latin America and the Caribbean (LAC)

The majority of countries in this region are now classed as lower or upper middle income countries, with one of them, Haiti, classed as an LDC (see Figure 4.5). But wealth does not mean it is equally distributed, and the disparity between rich and poor is extremely evident on the continent, resulting in high crime rates. A large development focus within the LAC countries is trade-related, enhancing national competitiveness in key export sectors such as agriculture, textiles and services. The region receives $10.812 billion in ODA (2010 data) representing approximately *8.2 per cent of total ODA.*

REFLECTIONS ON SOUTH–SOUTH DEVELOPMENT

The number of people from the global South working in development is increasing, which I believe is a good move. Being from a *developing* country yourself, gives you the advantage of greater insight into the problems you are trying to tackle and perhaps more respect from colleagues. It also has the advantage that you can blend in more easily. I first realised this when I was sent on an assignment in Columbia to work with internally displaced people (IDPs). Why me, I had no experience with IDPs? At a time with high risk of kidnapping, a blue-eyed, blond person would have been an obvious target; I could pass as a local if I didn't speak too much. In Indonesia my driver said I looked just like his brother. In Africa, I am seen as an Obama. This is an advantage, because to work well with people, you have to gain their trust, be at their level, have familiarity. When I worked in Afghanistan, everyone was looking at me, as I was dressed differently. So I started adopting the local dress code (tanbān, ezār) which seemed to break down the barriers between myself and the local team. Suddenly people spoke to me in Dari, which of course I didn't understand; however, we became friends and worked better together. It is these barriers, and misunderstanding, that make development projects fail.

In Afghanistan, I was brought in to resolve a failing project, to build clinics in the north of the country. A subcontractor had brought in their own masons from Australia, completely bypassing the local people, because they didn't trust them. But the local people happened to be a team of very qualified architects and engineers, trained in the university of Kabul, who were given no

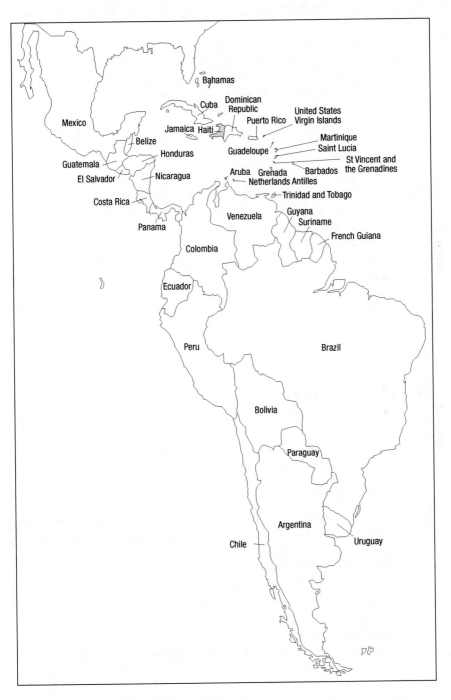

FIGURE 4.5 Latin America and the Caribbean map.

authority. When they were put in charge, they did a much better job than the Australian team.

Unfortunately, international staff members are often inclined not to trust locals, frequently perceived as inefficient, lazy and corrupt. But there are always some good and capable people out there. And that is my philosophy when I recruit – find the good local people, who can shine.

Milton (Honduras)

Europe

Europe received less than *4.5 per cent of the total world ODA* in 2010 and of the 49 countries in Europe only 11 received aid. These were Albania, Belarus, Bosnia-Herzegovina, Croatia, Kosovo, Macedonia FYR, Moldova, Montenegro, Serbia, Turkey (both Europe and Asia but taken as Europe by the OECD-DAC) and Ukraine. Top ODA recipients in Europe were Turkey ($1,049 million), Serbia ($651 million) and Ukraine ($624 million). However, net ODA received as a percentage GNI only exceeded 2 per cent of GNI in five countries: Kosovo (10.9 per cent), Moldova (7.5 per cent), Bosnia-Herzegovina (3 per cent), Albania (2.9 per cent) and Macedonia FYR (2.1 per cent). All countries receiving ODA in Europe were part of communist East Europe during the cold war years. None of these countries falls within the LDC groupings.

Pacific Islands (Oceania)

There are four LDCs in the region, which are Kiribati, Solomon Islands, Tuvalu and Vanuatu (see Figure 4.6).

DID YOU KNOW?

- Papua New Guinea is the country that is home to the most languages, over 700 in total. The most commonly spoken languages are Motu and Pidgin English.
- Oceania has a food-related problem but not due to insufficient supply. Quite the opposite: obesity. Several countries in Oceania have the highest obesity levels in the world. For example, for men 15 to 100 years of age, the top four countries with the highest rates of obesity are in Oceania: Nauru (84 per cent), Cook Islands (72 per cent), Federated States of Micronesia (69 per cent) and Tonga (64 per cent). These are all higher rates of obesity than in the United States, which at 44 per cent is in fifth position worldwide.

Data from World Health Organization WHO Global Infobase

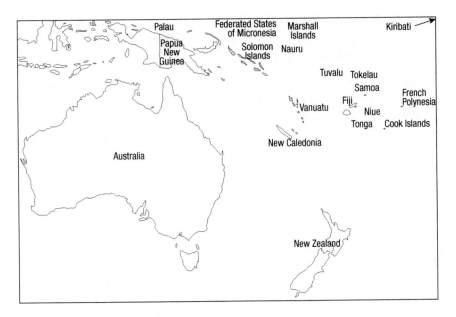

FIGURE 4.6 Pacific Islands (Oceania) map.

The total population of these islands is very small when considered on a world scale, being only about 9 million people. Consequently, ODA per capita is the highest in the world (averaging $221 per person) – and is extremely important to the countries of this region – despite the absolute numbers of ODA being relatively small (ODA receipts for the region being $2,019 million in 2010, only about *1.5 per cent of total worldwide ODA* for that year). Net ODA received makes up a significant part of the GNI of many countries in Oceania: Solomon Islands

---

HEAD OFFICE OR FIELD OFFICE?

I feel that sometimes the opinions and experiences of field-based staff are not given enough weight by headquarter colleagues.

Anonymous

Many of the larger development organisations will have their head office located in a donor country and field offices in recipient countries. This can sometimes create difficult dynamics, especially where headquarters staff need to make decisions about what is to be implemented and the priorities on the ground. In order to overcome these challenges and ensure that decisions made at the top represent real issues on the ground, communication between the two is vital, with head office staff that have substantial field experience.

CAREER CLINIC: SHOULD I DEVELOP COUNTRY OR REGION SPECIFIC EXPERIENCE?

Most people don't limit themselves to an individual country, instead opting to develop broader regional expertise and knowledge. Unless you have strong ties to a country it is likely that you will move around more as opportunities emerge. Donor trends and funding change, so by maintaining a wider focus (e.g. specialising in the Sahel region, Great Lakes Region, Central America or East Asia), employment becomes easier. In order to do this you might want to learn the common language in the region, understand political and cross-border relations in more depth, cultural particularities and so on, in order to give you a competitive advantage when applying for jobs in this region.

(61.4 per cent), Marshall Islands (48.6 per cent), Federated States of Micronesia (40.5 per cent), Tuvalu (34.7 per cent), Samoa (27.1 per cent), Palau (19.5 per cent), Tonga (19.5 per cent) and Vanuatu (15.2 per cent). Australia's September 2013 announcement to reduce aid will have a significant impact on the region and put more pressure on New Zealand to fill the gaps.

Working in a humanitarian emergency

Working in a humanitarian relief context is very different from a development context. Whether it's a famine, earthquake, flooding or insurgence, it's difficult to know where the next crisis will be. In the immediate aftermath of a disaster the environment is likely to be chaotic, fast paced, with little time to think or reflect. Whether it's the Asian tsunami and deployment to Banda Aceh or Pakistan floods, it is difficult to predict where the next deployment will be. However, funding for this remains relatively small, but protracted crises such as in Sudan or occupied Palestinian territories (OPT) receive the most significant levels of funding.

Living conditions can be very basic – a camp bed if you are lucky – and there may be very tight security restrictions. The work is intense and your team members become very important as you work and live together, and turn to each other to make sense of what is going on around you. Organisations select teams and recruit people very carefully; you can't have an unstable, overly sensitive or needy person in this type of environment. Only experienced people tend to be recruited for this type of work.

As the situation stabilises, some organisations stay on to support reconstruction efforts. Here, people with less experience may be recruited. During longer term reconstruction work, or in situations of protracted crises, the work bridges development and humanitarian relief.

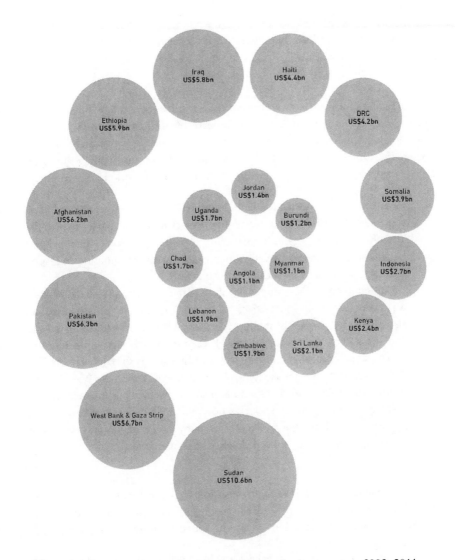

FIGURE 4.7 Top 20 recipients of international humanitarian response, 2002–2011.

Source: Reproduced with Permission from Global Humanitarian Assistance, www.globalhumanitarian assistance.org

I was M&E Manager for an NGO in South Sudan. I had to travel a lot to oversee data collection and train staff. I usually stayed in one of the staff camps. My room would be an army tent (although sometimes a mud hut) and there was a shared dining area. I had access to an office, but no desk of my own, so I had to hot desk all the time. I was out in the field most of the day and came back in the evenings to emails and often urgent proposal deadlines. Between areas I travelled on small planes. I moved on after a year, partly due to the hard living conditions, but I worked with some great people – both local and international staff.

Mya (UK)

The location of humanitarian work is of course still determined by flows of donor funds, directed by the latest crisis or long-term reconstruction efforts. Inevitably, many are concentrated in poorer countries, where there is less investment in disaster preparedness and local capacity to deal with the situations. Inevitably, security risks can be many and varied, which creates challenges in the working environment.

I was Finance and Admin Manager for an NGO in Gaza. At a technical level it was straightforward but in every other aspect it was extremely challenging. The office moved twice in three months for security reasons – being close to beneficiaries is a mixed blessing when Israeli troops are choosing to respond to attacks from Gaza by shelling civilian areas.

Although my role was nominally office based, in reality all staff did whatever was needed to achieve objectives, so I sometimes did HR, procurement, and also established a food distribution operation. All of us survived on limited sleep and stress levels were high, but my NGO gave us R&R every three months.

The situation between Palestinians and Israelis was so highly charged that it was often difficult to make things happen, and it was easy to become angry or frustrated with individuals who appear obstructive, but in reality most of them are scared and don't know how else to behave. It is a man-made humanitarian crisis but very different from my later experience in Haiti, where the main threats were random violence, rape and kidnap.

Phil (UK)

Not all humanitarian work is so highly charged. In many situations, it's long term, and is less about dealing directly with the crises. The focus has shifted more towards rehabilitation and reconstruction, laying the foundations for development.

I found it very hard to reconcile in my head the two worlds that I was seeing in Haiti. During the day I was out in rural hospitals collecting data on the cholera epidemic and meeting very poor people who had lost everything. Juxtaposed with the people living in shanty towns in Port au Prince were massive mansions for the UN staff and other international agencies, with their big swimming pools, and wine and champagne parties at the weekend. It is very difficult to come to terms with this. I sometimes felt rather ill at the money that was being thrown around on the staff of the big agencies, people who weren't always very effective in their jobs.

Anonymous

CAREER CLINIC? ARE YOU MORE SUITED TO EMERGENCY OR DEVELOPMENT WORK?

The constant adjustment linked to travelling to and from different emergency situations and the high workload can make this work quite stressful and demanding.

Annelies (Norway)

This really depends on what kind of person you are. Here are some questions to make you think about this:

- Do you enjoy working in the midst of chaos alongside 'strong personalities' (emergencies), or do you prefer to work cooperatively and be able to anticipate what is likely to happen and plan for it (development)?
- What's your personal 'risk threshold' like, and are you happy to work in a conflict setting?
- What kind of lifestyle do you want and what are you hoping for in your personal life, as emergency work isn't always compatible with sustaining long-term/long-distance relationships or having a family.

If you have emergencies training and would like to put it into practice, get field-based emergency experience in early on, when you have fewer commitments etc., and then move into development work, but if you've developed a thematic focus it should allow you to move relatively easily between the two.

Notes

1 While aid has increased by 63 per cent over the past decade, the 2012–2015 survey estimates that the donors' aid volumes will decrease by 0.5 per cent per year from 2012 to 2015. *2012 DAC Report on Aid Predictability: Survey on Donors' Forward Spending Plans 2012–2015*, OECD, Paris, available from www.oecd.org/dac/aid-architecture/2012_DAC_Report_on_Aid_Predictability.pdf (accessed 30 January 2015).

2 Many so-called *donor darlings* still face funding gaps in absolute terms, making rebalancing from darlings to under-aided countries within any given global resource pool undesirable as well as impractical.

3 *Identification and Monitoring of Potentially Under-aided Countries*, OECD report (2013).

4 Compiled by the Committee for Development, a subsidiary body of the UN Economic and Social Council. The distinction first arose in 1971 in the United Nations Committee for Development Planning quest to examine challenges peculiar to the least developed among developing countries.

5 *Identification and Monitoring of Potentially Under-aided Countries*, OECD report, (2013).

6 *Aid to Africa*, Policy brief no. 1 – October 2010 available from www.un.org/africa/osaa/reports/2010_Aidbrief.pdf (accessed 30 January 2015).

7 In fact, Afghanistan was the top recipient of ODA in the world, receiving approximately 5 per cent of all ODA in 2010.

8 World Bank review of MENA region.

PLATE 3 Chuon Chhon spent the first eight years of his life confined to his parents' house crawling on his front like an animal. The community at the time used the word 'snake' to describe him. After years of home based rehabilitation through the Cambodian organisation CABDICO, Chhon is now an active and productive family member.

Source: Anthony Jacopucci/Handicap International. Reproduced with permission from CABDICO

PART 2

Breaking into the sector

The first four chapters have given you an insight into what it is like to work in the sectors of development and humanitarian assistance. This leads us on to the next big question – how can you secure your first job in the field? Whether you are starting out in your career, or making the transition from another area, finding your first development or humanitarian role can be a complex process, navigating through unknown territory. The next four chapters are here to help you answer this question and gain some clarity.

- Chapter 5: Routes in will help you to understand the entry routes for some of those currently working in the sector, as well the broad characteristics of the sectors and what you can do to enhance your employability within them.
- Chapter 6: Academic qualifications primarily focuses on Master's courses, whether you should do one, how to select the right one for you, and making the most of your experience as a student. Undergraduate and doctoral courses are also covered briefly.
- Chapter 7: Volunteering and internships is for anyone considering doing some volunteer work – whether to overcome the catch-22 of no-job-without-experience or for more experiential or altruistic motives.
- Chapter 8: The job search is meant to be a companion to help you find and secure your ideal job in the field. There is no magic bullet for this, and the chapter highlights the many different strategies at your disposal and troubleshooting if the search is not progressing well.

5

ROUTES IN

In recent decades the development and humanitarian sectors have professionalised – a much welcomed move. Along with this has been the birth of the development 'industry' and the shift for development organisations, including NGOs, to function more like private businesses, with their expenditure and value for money coming under tighter scrutiny. What does this mean for the job seeker? Private sector experience is highly valued as is the need for professionalism and marketable skills. The challenge is that with very few structured graduate training schemes, you will need to take a much more active role in planning your entry into and career in the sector.

The first section of this chapter will help you to understand how some made it into the sector, and why some didn't succeed, while the second part looks in more detail at some broad characteristics of the sector and what you can do to enhance your employability within the sector.

What is your story?

Each person in the sector has their own unique story of how they got their first paid position, and how they used this to propel them forward in their career. There is no typical or direct route in, but this chapter serves to give some examples of how some people have moved into the sector and some of the more common routes.

> I am a roving nutritionist for GOAL and work to improve food and nutrition programmes for some of the poorest people imaginable. It's a job that I love, but in common with many aid workers, it took me some time to realise that it was what I wanted to do. I graduated from university with a degree in Biomedical Sciences. Career counsellors told me that my profile matched that

of a prison warden, which was not exactly what I had in mind! Various PR, media and marketing jobs left me underwhelmed, so I headed overseas and found myself in Phnom Penh (Cambodia) working for a local NGO. I also taught English in a city slum orphanage. This rich experience provided me with focus and purpose, and upon my return home I enrolled with a Public Health Nutrition Master's programme at the London School of Hygiene and Tropical Medicine (LSHTM). Following a gruelling year studying a range of topics from epidemiology to nutrition in emergencies, I completed the course. The toughest challenge still lay ahead: landing my first job. I sent out numerous applications and finally, three months later, I got my first break and found myself in Burundi. Since them I have worked for a number of organisations, mostly in Africa and Asia.

Hatty (UK)

My gap year in Tanzania was what really motivated me to work in development. I studied Modern Languages at university and then went straight on to a Master's in Development Studies. It was a long process to get my first job, entry positions were very competitive. But persistence paid off; after two years of applications I got a paid internship position in a small organisation. I did programme support as well as some development education work in schools. Through this I got involved in the Make Poverty History campaign in 2005. It was an exciting time, and exposed me to the importance of campaigning. As I built up experience in this area, I was then able to apply for advocacy and campaigning roles and over the last seven years have worked in this field for three different organisations. Over this time I have had the opportunity to work with incredibly inspiring people across the world who have made me passionate about the difference that advocacy and campaigning can make to poverty and injustice.

Celine (UK)

Where I come from in Northern Ghana, there aren't too many opportunities for employment, but there are a lot of NGOs! At school I got involved in some voluntary work with one NGO, doing health outreach training with local communities. I enjoyed the work a lot, especially as I was helping my community. Young people here don't often see the point of voluntary work as it is not paid, but for me it proved to be an important route in. After I graduated from university in Accra, I returned home and (as I was jobless) continued to volunteer with the NGO. Eventually they offered me a job as a programme assistant, then coordinator. I am now their West Africa regional coordinator managing health and education programmes.

Moomin (Ghana)

I started out in the sector quite by accident. I was a civil engineer, working in my home country of Honduras, when Hurricane Mitch struck. All of a sudden, rich, poor, everyone was equally affected: houses were flooded, roads destroyed, we had to all go down the road to collect drinking water from the tap. And then the aid and NGOs streamed in. As I had just completed a Master's in Urban Planning, I decided to start attending weekly municipality meetings, which convened all the NGOs. I listened to people from around the world who knew very little about the country discuss some of their reconstruction projects, and what they were going to do. Officials didn't really question, or give direction, as these organisations 'maybe knew better' and came in with money. But some of the ideas were terrible, as they didn't know the local environment. I started to facilitate, giving my opinion, making suggestions for improvements, buying maps to illustrate what I thought should be done, etc. And people started listening to me. One of these people was the Country Director for Global Communities (formerly CHF International). We stayed in contact and when my current contract came to an end, she offered me a job. Since then I have worked with CHF in Columbia, Bolivia, Afghanistan, Cuba, Rwanda . . .

Milton (Honduras)

After graduating as a doctor, I felt I didn't fit in to the stereotypical medical career; I wasn't an academic wanting to make my way up the ranks. So I did a short course in humanitarian emergencies and took up my first short-term role in the DRC. I haven't looked back: Afghanistan, Haiti and Somalia are some of the places I have worked in.

Tom (Canada)

I studied Accountancy but after graduating I was looking for an interesting career that would allow me to travel. Someone mentioned humanitarian work to me, which sounded interesting, but I felt I needed some private sector experience first. After two years I then applied to work as an accountant with an international humanitarian organisation in Afghanistan, and was selected, as accountancy is a skill that is in demand. I have since worked in over six countries in Africa, Asia and Latin America with four different organisations.

Amitava (India)

I knew I wanted to work in development from a young age, inspired by my father who was a journalist in South America and Africa reporting on the newly independent countries. I did an undergraduate degree in Development Studies, followed by a Master's in Human Rights and then another Master's degree in Development Studies, for which I did six months of fieldwork in sub-Saharan Africa. Several years of volunteer work and experience in the civil service led me to apply to the UN Junior Professional Officer Programme and persistence paid off. I was successful and on my fifth time of applying I was accepted, just before my thirtieth birthday. I am now working for the UN, currently stationed in East Africa.

Bjorn (Sweden)

I was working in advertising in London but always had a penchant for development work and a special interest in Africa. Finding the right role was tricky as my background is very specifically brand and advertising-related – an area that is relatively in its infancy within the development world. I spoke to my headhunter about my planned move who was supportive. It was the one route I tried, and the one route I was lucky enough to succeed in. He suggested that I apply for a role in brand-led development with an organisation that stemmed from a Public Private Partnership (PPP) empowering girls. The role suited me perfectly. After many interviews, Skype calls and emails, and a visit, the rest is history. Now I live in Rwanda driving forward a brand that is helping create positive social change in the country, working alongside other NGOs, closely with the government and with our counterparts in other countries.

Richard (UK)

I became interested in working in development, but with a career as a theatre and set designer, a husband and small children, it was a difficult transition to make. Instead, I set up my own company at home, which was very successful. After my children had grown up, I did a degree in Development to understand the issues in more depth and what my role might be. At my age, I didn't want to get an entry level job with an NGO, so I decided to set up my own fund. I now work with energy efficient technology, installing stoves in both households and institutions. Long-term sustainability lies at the core of our work and our process is focused around local ownership through strong partnerships with all stakeholders.

Line (Norway)

Having been an accountant in the oil industry I decided I wanted to move across to development and NGO work. While the skills are similar the context is very different and initially some organisations rejected me as I didn't have enough development sector experience. A small NGO was happy to take my experience and skills, and although I worked hard and contributed a great deal, I still believe I gained much more from the opportunity. Rather than being in a specialist finance role I became involved in logistics, admin and programming as well and my flexibility was a real asset so I learned a lot in a few months. I would advise anyone aspiring to work in a field-based NGO to be flexible and adaptable, as they will learn much more from this.

Phil (UK)

In Part 4 you will find many more stories from people about how they made an entry into their areas of specialism and moved up in their careers.

But not every story is a success story. It's also relevant to learn from those who have tried for a career in development or humanitarian work, but have not been successful for various reasons. Here are just some examples (Chapter 12 also talks about why some people chose to move away).

I did my undergraduate degree in Development Studies, with field work in Papua New Guinea. After I graduated I applied for a large number of entry level jobs with NGOs but was not successful. I finally accepted a position as an office manager for a large company in Perth, telling myself that I would gain some experience in the private sector for a few years, and then try again. Six years went by before I got the drive to start applying for development jobs again, but my lack of international experience seemed to be a problem. I finally decided to take a six-month career break, and became a Kiva-fellow, working with a microfinance organisation. You pay for the experience but it is well organised and I learned a lot. When I returned home I started applying for jobs once again, but didn't get anywhere. I was offered my old job back, so I took it as I needed the money.

Catherine, (Australia)

I worked in banking but loved travelling and wanted to combine the two, as well as do a career that I found more fulfilling. After I had eventually saved up enough money, I quit my job, and did six months of volunteering with a small NGO in Guatemala. While I was working to support some women in an income generating activity, I was a bit out of my depth. I didn't know the context, my suggestions for change weren't eventually relevant and I didn't speak the language well. After I left I realised that my support didn't have any long-term effect. I thought about doing an evening course, or part-time Master's in Development but I didn't find the time. The experience did however help me to secure a job back in London as a fundraising officer for a local NGO. They weren't international but I told myself this was the first step to gain experience. At the age of 34, I found the work very menial, boring, had a bad boss, and worst of all was getting about a fifth of my salary back at the bank. I decided my lifestyle was more important to me, so I went back into banking.

Lisa (UK)

Some characteristics of the sectors

What can we learn from the success stories? And from those who eventually choose a different route? First, let's look at some of the things you need to consider about the humanitarian and development sector, which will help you to get on to a good grounding.

Good intentions are not enough, the sector needs specific skills

We see that personnel with technical and engineering skills are in high demand in international emergency operations. In Norway, at least, people looking for careers in development and humanitarian assistance tend to forget about this option and instead choose to study political science etc. where the competition for jobs is much higher.

Benedicte Giæver, Director of NRC's Emergency Response
Department, which operates the emergency
response roster NORCAP

In the past, the development and humanitarian sector – in particular NGOs – was defined by charitable donations and volunteerism. Simply 'giving your time' was a good and noble cause. This may still be the case if you are doing volunteer work in your own community, but if you are looking for a paid position, you need to be able to offer marketable skills that are not locally available. What are the skills that you are able to offer? Sometimes these will be technical skills, other times project coordination skills, writing skills or a proactive work ethic. Be clear about what you can and will offer.

Social science majors and liberal arts students need not be discouraged. I was a history and religion major, which on the surface seems rather unhelpful. However, the skills that got my foot in the door were writing and research. Moreover, projects and programmes need the support of social research to measure their impact and appropriateness. Such tasks require good social scientists.

Matthew Bolton (USA), Former Humanitarian

Cross-cultural understanding is vital

While technical knowledge is extremely useful, contextual knowledge also has a vital role to play. Too much 'bad development' has happened through ignorance and naivety. Many issues are complex and deep rooted and may not be what immediately meets the eye.

In order to get a job done in a culture context that is not your own, you need to have an understanding of the local customs, politics, priorities, ways of working and how to navigate complex power-infused relationships. Flexibility and adaptability are two traits that will stand you in good stead, as will a recognition that your way of doing things is not always the best – in that environment at least.

This type of understanding and open-mindedness comes only from experience, no amount of textbook study or technical knowledge can prepare you to work across environments. For this reason the sector places so much emphasis on international experience: living and working in different contexts, being open and sensitive to cultural subtleties.

One of the challenges of working cross-culturally is that people often have different expectations. There is a disconnect between my reality, my agenda and your reality. This creates a scope for misunderstanding and conflict that adds an extra dimension and challenge to the work that you wouldn't get to the same extent back in your home country. First, you have to understand how other people interpret things, and then try to balance different views. When it comes to evaluation, how are we actually measuring the success of a project when different people involved have different agendas?

Mya (UK)

Persistence pays off

While obtaining a job in [development] is by no means a straightforward undertaking, if you feel passionately about working in this field there are fulfilling ways to get there! This will involve creating the building blocks of your career through a careful mix of postgraduate training, experience, skills and net-working. Research the area in which you want to work, consider what kind of work you want to undertake and work backwards to plan milestones and your immediate next steps. You may well also redefine and refocus along the way.

Oxford University Career Service

There might be 200 applications for one job. It's a harsh reality, but it is often the case, especially for entry and mid-level jobs. Getting noticed isn't easy, and it involves a mix of luck, ensuring that your experience makes you stand out, networking and utilising all the available avenues to get your first position. Once in, your first job may not be your ideal job, and it is likely you will have to make multiple strategic job changes in order to get where you want to go.

This is without mentioning low salaries, and potentially the need to do unpaid volunteer work and invest in your own education in order to gain the skills that will put you at the same level of 100 other candidates. Sound tough? If you are discouraged at this initial hurdle, the sectors may not be for you. Once in the job, the reality you encounter to achieve *development outcomes* or work in a humanitarian context can be much harsher, so those weeded out at the start may not actually be cut out for a career in the field. Persistence, coupled with informed and intelligent choices do usually pay off.

Your employability depends on your own initiative

Development is a fast-changing and fluid sector, constantly evolving, with new skills being in demand as trends or development fashions emerge and change. Planning your career carefully, making informed decisions and keeping up to date with the latest development and investing in your own professional development will ensure that your skills are marketable and responsive to the demands of the sector.

Anyone wanting to work in the field needs to know that Africa and other continents for that matter are changing fast, as are aid and donor policies.

We have to change our perception of – and attitude towards – developing countries. At this stage of my professional career, I focus on making markets work, with the right mix of private and public sector actors and a sidelined role of development programmes. The agribusiness sector is one of the most complex sectors that need the collaboration of many stakeholders. But to make the private sector work well we still need external support which, if based on dialogue and institutional embedding, can still make an important contribution to sustainable development.

Ted (Netherlands)

CAREER CLINIC! THE IMPORTANCE OF FOCUS

You will have to have a much more specific goal than 'do humanitarian work'. There are dozens of options within this; do you want to be a logistics coordinator, do humanitarian response management, develop curricula and training materials for primary education, be a nutritionist, work with refugees, or might measuring development outcomes be your thing? Indeed, many people's careers are quite messy, and they are likely to assume several quite distinct roles throughout their lifetime. This is inevitable, but when planning your career, focus will help you to stand out. If you know where you want to go, you are more likely to get there.

It can be difficult, with little knowledge of the sector, to understand what you may be best suited to. Reading, research, some work experience, speaking to people in the field will be a good starting point to help you start to determine what types of jobs and roles attract you the most. Part 4 of this book helps you to understand the breadth of possibilities that exist.

Having a clearer direction will help you to shape all your studies, volunteering and work experience in this direction. The more focused and relevant to a particular role or organisation, the greater your chances of standing out to the recruiting manager when they are shortlisting candidates for a role. Find jobs you would like to be doing, meet people who are doing those jobs, and understand how they got there.

Ways in

Listed below are a number of different strategies that may help you to break into the sector. Evaluate where you are currently at, and what steps you need to take to enter the sector. Do you need to gain more experience, and thus do some international internship, or is it skills or academic studies that you are missing? In isolation none of them is likely to be sufficient, and you will need to determine which combination will work best for you, based on your current starting point.

Volunteering/Internship

From a few weeks, to several years, there is a huge range of options out there. In some positions you may be paid, covering your own costs or even paying for the experience. With the right organisation, this can be a good way to build up skills and contacts. If you are overseas, it can also be an invaluable way to build up international experience. But do weigh up the opportunity costs of working for no or a low wage, with the potential benefit to your career.

Chapter 7, Volunteering and internships, takes you through this option in more detail.

Academic qualifications

A Master's degree has become a tick-box criterion even for many entry jobs but is by no means always a requirement if you have enough experience to make up for it. You may be thinking of doing a Master's in Development Studies or a related area to launch your career in the field. But there are many options, and some may provide more direct entry routes than others. It's worth exploring the options carefully and choosing the right one at the right time for you.

Chapter 6, Academic qualifications, will take you through this in more detail.

A training scheme

A limited number of organisations offer graduate training schemes. Responding to demand, some organisations are launching such schemes, while others, seeing the high volumes of over-experienced applicants, are discontinuing them, preferring to recruit people directly into jobs. While you are rarely guaranteed a job after the scheme, it can be a great way to launch your career in the field.

Appendix 2 lists some of the organisations offering training schemes and their details.

Learning on the job

By working your contacts, or by sheer luck and being in the right place at the right time, you may find yourself relatively inexperienced, but in a paid position within the sector. Congratulations! The challenge now will be how to develop in your career, and plan your next move. One paid role does not always mean a smooth transition on to a next position. Understand how you can develop in your career.

Chapter 8, The job search, takes you through the various different ways you can find your ideal job.

Chapter 9, Advancing in your career, helps you to plan and prepare for your next move.

Gaining skills in another sector

Private sector experience is increasingly in demand in the NGO/public sector. If you are just starting out, the private sector can offer structured training and personal development opportunities, which can help you to gain skills that can be transferable – and are often very much in demand – in a development or humanitarian context. While changing sector is not always easy, if your skills are in demand and you plan your transition carefully, gaining relevant overseas experience for example, it can be a good way in.

> My daughter studied maths at university but decided after graduating she wanted to work in development. Several unsuccessful job applications later, she identified the lack of international experience as a bottleneck and went off for four months to Somaliland to volunteer (at great expense to me). She came back and tried her luck at jobs once again, and once again failed. The sector needs specific skills, and she had none to offer! With reluctance I persuaded her to apply for an accountancy training scheme in London. Accepted by a blue chip company, once she qualifies as an accountant, she will be in a much better position to offer her skills to the development sector – if she can accept the salary cut!
>
> *Disgruntled Dad, London*

Starting your own NGO

Some people may feel so passionate about an issue that they consider starting their own organisation in order to address it. This is no easy undertaking and it often takes considerable energy with very little financial return in order to get off the ground. NGOs started merely to address one's own visions are rarely successful, but if you have the passion, drive and entrepreneurial spirit to see your idea through, you may be able to make a success of it.

Chapter 11, Starting your own NGO, will help you to think through this option further.

Selling your services

Many organisations often hire short-term consultants, with specific skills, to undertake specific pieces of work. This may be a piece of research, evaluation, producing a manual or delivering some training, for example. If you have specific skills, doing consultancy work in the field can be a good way to get experience and contacts within a range of organisations.

Chapter 10, Working as a consultant, explains some the particularities.

> My interest was in gender and I did my Master's in this area. To gain more experience I decided to join VSO for one year, and was a gender advisor in a

government department in Central Africa, which was a great experience and gave me a lot of credibility. I realised many organisations were gender blind – often because they did not have an in-house gender specialist. I set my own small consultancy firm up and now advise organisations on how they can mainstream gender into their programmes. I see myself as an agent of change on the ground, slowly changing people's attitudes and emphasising the importance of including gender in programming. I get a lot of work from local NGOs, INGOs, donors who review their strategic plans.

Anonymous

CAREER CLINIC! HOW USEFUL IS TIME SPENT ABROAD IN A TEACHING ROLE TO LAUNCH MY CAREER IN DEVELOPMENT?

Having international experience is incredibly valuable if not imperative to get a good job in the sector. If you want to immerse yourself in a country but haven't yet built up any specific marketable skills relevant for volunteering in NGOs, or need to earn some money, doing a teaching placement can be a good way to do this. The demand for English and science teachers is high. There are opportunities to get involved in classrooms without formal teaching qualifications but if you want to teach English, a good long-term investment is to do a Teaching English as a Foreign Language (TEFL) qualification, which will allow you to be employed overseas (and cover your expenses while exploring the field further). Many development professionals launched their careers in the classroom. Being in the country helps you to build networks, improve your language skills and learn things that you wouldn't get from academic texts. In your spare time, try to get involved with local NGOs in your area of interest to further boost your experience and CV. Two organisations, Tenteleni (www.tenteleni.org) and www.vesl.org, place volunteer English as a Second Language (ESL) teachers abroad.

6

ACADEMIC QUALIFICATIONS

Formal education provides an important foundation for a career in development and humanitarian assistance. This chapter primarily focuses on Master's courses, which have become an industry standard. Regardless of your background, a Master's can provide an entry point into the sector, or an opportunity to specialise and advance in your career.

The chapter is divided into four sections. The first briefly looks at relevant undergraduate degrees that can set you up for a career in development or humanitarian assistance. The second covers some important things to think about before embarking on a Master's course. The third part of the chapter is for you, the current student. How can you make the most of your time as a student, and make it count towards your future employment? The fourth section briefly covers PhDs. A selection of short-term or professional courses aimed at mid-career professionals can be found in Appendix 2 Continuous Professional Development.

Undergraduate courses

Many people come to development or humanitarian assistance later on in life, after they have already completed their undergraduate studies. However, if you know this is the area you want to work in early on, you have three choices in terms of undergraduate degrees:

1   A professional/technical degree. A degree that will give you a profession, which is also very much in demand within the development/humanitarian sectors. This includes anything from the health sciences (e.g. medicine, nursing, public health), engineering, law, water management, economics and public finance, architecture, etc. You will need to build up a solid foundation, with some

years' experience in your home country before you can start to make the transition to international contexts. Gaining international experience while a student, or early on in your career, through volunteer placements for example, will ease your transition.

2   A social sciences degree, such as political sciences, international relations, anthropology, geography or development studies. These types of degree provide an important knowledge foundation for working in the sector but again, you will need to gain experience and some specific marketable skills.

3   If neither of the above options are for you, don't worry. Focus on something you are passionate about and interested in. You will also come across people with all types of academic backgrounds who have made a career in the sector: biology, sports, art and design, astrophysics or modern languages.

> I decided to do an undergraduate degree in development studies because I thought I could change the world. In my first year I realised that this was an illusion. Not only this, but that the organisations that I had previously thought highly of – NGOs, UN, IMF – were not always good (we also did a lot of criticism of their work in the first year). While studying an interdisciplinary subject such as this does have its drawbacks in that you have very broad knowledge, it does also have positive aspects. First, it gives you important perspectives on the historical and social background. Second, it helps you to make a more informed decision about what you may want to specialise in. One area I have become very interested in is Behavioural Economics, and I think I may go on to do a Master's in this area.
>
> *Katharina (Austria)*

Master's degrees – some important decisions

Master's courses are typically one or two years of full-time commitment – or up to five years as part-time studies. Given the financial and time commitments, most people will only do one full-time Master's in their lifetime so it is important to make decisions carefully. Having said this, it is not at all uncommon to find people with two or even three Master's – done part time while working – to progressively gain new skills in line with the changing demands of the sector. The question is when is the right time to do it, what should you do it in, and where?

When is the right time to study?

> I wish I hadn't been so impatient to work in the development sector. After my undergraduate degree I went straight on to a Master's in International Development. It was the most obvious route into the field. After graduating I found it hard to get a job and had to do several volunteer and short-term positions. If I had done this *before* my Master's I would have been able to get

a greater insight into the sector, gain more specific direction, and use my Master's as an opportunity to specialise in the specific area that was of interest to me.

Anonymous

Should you go on to do a Master's immediately after an undergraduate degree, or later on in your career? It is an advantage – and often strongly advisable – to get a few years' experience first. Indeed, many admissions committees give preference to applicants who have some years of (ideally development) work experience, as it enables you to contextualise your learning and make a more informed contribution to discussions.

Fieldwork can clarify your preferred line of expertise and will allow you to focus on your specific areas of interest. Once you graduate you will offer employers an increased level of maturity, experience and knowledge about the field, which will enhance your chances of being recruited.

There is a catch-22 here – how do you get a job without experience or a Master's? Be creative, target smaller organisations, and use your networks. Chapter 8 will give you some ideas.

If you pursue a Master's once you have more experience, particularly international experience, it will make it that much easier to get good quality internships while you are studying – and that can be just as important as the qualification itself.

Jeff Riley, Careers Consultant at Queen Mary, University of London

You can read more from Jeff about choosing a Master's at www.gradsintocareers. co.uk/development.aspx

What to study?

For some employers, having a Master's is just a tick-box criterion. They may be less focused on the actual content of the course, looking instead for good grades, initiative and ability to learn on the job. However, a Master's can also open up many doors, and expose you to skills, knowledge, contacts, networks and opportunities in that field. You should do a Master's degree for these reasons, not for the paper qualification.

Having a clear idea of what you want to do and where you want to go is the first step in identifying the right Master's for you. A Master's course that specifically relates to the job or field you want to work in will, of course, give you a definite competitive advantage over other candidates with more general degrees. Plus it will allow you to shape all your experiences in this direction and get you the job. If you are not yet sure where your interests lie, or have yet to narrow them down, there are many different factors you have to weigh up. Here are some questions you need to ask yourself:

- What inspires, drives and interests you?
- What is your ideal job? (If you can't yet answer this some work experience or internships may help.)
- What are the common qualifications for your ideal job?
- What relevant skills do you currently have and what studies might complement these?
- What are emerging areas in the field, current skills gaps, and what are employers looking for?
- Do you want to gain an overview of the field, or an in-depth understanding of a particular development theme, or be a technical expert?
- Are you more attracted to management or do you prefer specialist work?

MA IN INTERNATIONAL DEVELOPMENT?

Logically it would follow that if you want to work in International Development, a Master's in the field will set you up to do this. A decade ago an International Development Master's may have helped you to stand out; these days it is not the case. While understanding the history, academic theories and different development practices is extremely important for anyone working in the field, it does not provide many of the *skills* required *to work* in the field. If you do not yet have marketable skills, gain these first and then go on to do a Master's in International Development later. Alternatively, there are several courses that combine a specific area of expertise (e.g. agricultural economics, gender, education, project management) with the development angle. As the joke goes, a development student asks her tutor how she could use her degree to get into development work. He replies 'Oh no you can't, people in developing countries need those with real skills such as doctors and engineers, not just another person with an arts degree telling them what to do.'

Student perspectives

I first did a postgraduate diploma in Development Studies, and stopped there as I didn't know what area I wanted to do research for the dissertation on. I then got a job with an Irish INGO on their governance and human rights programmes. During those first few months in the field I identified a research area that was of interest to me and directly linked to my work. I was lucky because my school in Dublin was set up to facilitate distance learning and I was able to write the dissertation overseas. I then put the two together, the dissertation and the coursework from the diploma and was given a Master's. Six years later I did a second Master's (in Adult Education). After many years of work, this education experience felt different. It helped me to reflect on and deepen my practice and provided me with new direction.

Sive (Ireland)

Development studies doesn't lend itself to early specialisation. It is best to get experience – or ideally specialise in another field first – and bring purer knowledge with you into your work. I became interested in development at an early age, but fortunately there weren't many undergraduate courses in Development at the time, so I did Economics and Political Sciences and moved across once I had some more work experience.

Shandana (Fellow, IDS)

I did Geography as my undergraduate degree and wanted an internationally oriented career, so I went on to do a one-year Development Studies Master's programme. There were four compulsory modules, all very broad, and then four months' fieldwork for the thesis. After graduating I did not feel like a professional in any way, and didn't feel ready to enter the labour market. I didn't even know how I would be able to practically apply any of the knowledge gained. But I had worked hard, and that was visible to my teachers, one of whom helped me to get an internship at the Africa Department of the Dutch Ministry of Foreign Affairs and after that I got temporary contracts at the Netherlands–African Business Council (NABC).

After one-and-a-half years NABC proposed to send me to Ethiopia on a flexible contract (self-employed; taking projects from NABC), which I am doing right now. So in the long term a Master's did help me launch my career, but indirectly through contacts, rather than practically applicable skills.

Auke (Netherlands)

I completed the Master's in Humanitarian Action with the Network on Humanitarian Assistance (NOHA) Universities, which is a full-time course over 16 months and is delivered jointly by seven participating European universities. I chose NOHA because of its reputation for being 'Humanitarianism 101'. The great thing about this programme is that it caters to both absolute beginners in humanitarian work and those who have field experience. While the latter can benefit from refreshing their knowledge of topics such as international humanitarian law or being exposed to medical humanitarian theory, those new get to explore it from both an educational and practical viewpoint – most of my fellow students had two to three years' experience.

Anonymous

From the moment I walked 'out of the classroom and into the field', I knew the macroeconomic training I had would do nothing for me as I wrote proposals, developed monitoring indicators, navigated tricky, power-infused relationships, and submitted financial reports. At the end of the day, it was the project management and 'soft' skills that became the backbone of my daily life as an aid worker. And what a degree could not have prepared me for, I believe, was the creativity and ability needed to size up a difficult situation and offer a way forward. This leaves me wondering . . . are some aspects of being an effective aid worker just inherently inborn or learned only through the hard knocks of life?

Jennifer (USA)

I did an MBA when I first started working in the development sector. Prior to this I had done an engineering degree and urban planning, but working in an NGO I felt I needed a whole range of other skills I didn't have: financial management, HR, the legal framework, marketing, etc. These skills have been invaluable for me to move up with my career, now at a Country Director position.

Milton (Honduras)

I did Cambridge's MPhil Development Studies programme. My fellow students had substantial field experience: heads of NGOs, Iraq veterans and former diplomats. The degree was a means to increase their marketability and move up the career ladder (just as an MBA would be). I wasn't in this league, but it was the perfect degree for someone going into academia or diplomacy. My research is on state formation in Central Africa; studying economic development with Ha-Joon Chang. The degree certainly contributed to my career preparation. You might say I'd have been better off with the MPhil Economics, International Relations, Politics, etc. But none of these degrees has the same interdisciplinary perspective as Development Studies. The Economics degree more often resembles an advanced mathematics degree, while Politics and International Relations are often focused on research methods and Eurocentric theories of great power politics. The only postgraduate option for institutions, identities, and 'irrational' economics in Africa was the MPhil Development Studies. I couldn't be happier I pursued it.

Zach (USA)

I did a Master's in Health Promotion because it was a field I was very passionate about. I have now worked the field for four years and am thinking of going back to do another more general course in Development Studies. I think this will help me to gain a broader perspective and understanding of the sector that I am now looking for.

Alessia (Italy)

I am in the process of completing my third Master's, and each one is complementary and broadens my skills. I decided not to do a PhD because I felt it narrowed down to a particular field too much.

Anonymous

My gap year in Kenya gripped me. I went back as often as I could and had a vague notion that I might like to work 'in development', in a job that I found personally fulfilling. But I didn't really know what 'development' was. I got some career advice where I was advised to a Master's: the best advice I got. I did a part time Master's in Rural Development at Sussex University while continuing to work as a management consultant. It gave me the confidence to look for jobs in the field, and demonstrate my commitment and credibility to employers. Shortly after graduating I was offered a position in Kenya.

Nick (UK)

Where to study?

I spent a long time researching the right Master's course. I wanted general Development Studies for the broad academic overview since my Bachelor's was not related to development. But I also wanted a course that gave me the opportunity to specialise in something, something practical that could be useful for the job market. So I opted for a course with a strong research focus, with the possibility of an extended period in a developing country, and only one year as I did not want to be out of the job market for too long. I didn't want to study in my home country of Italy as I find the teaching not critical enough. With all these criteria I started looking around for a Master's and came up with the University of Amsterdam as the only possible solution. It combined all the criteria I wanted and was affordable.

Marcella (Italy)

Once you have a thematic focus, how do you choose a course and university? Here are some criteria you may want to use to help you to narrow down your choice:

Country: Doing a degree with an international focus, it can be an advantage to complete studies in a country other than your own. Being able to demonstrate that you can work in cross-cultural environments is an advantage. Gaining exposure to different cultures will also make you more open-minded, have a global outlook and understanding. Effectively, it can help you to 'kill two birds with one stone' as the saying goes.

As someone from the global north, studying in a developing country can help you to gain much needed international experience, at the same time as being close to volunteer opportunities with local NGOs. Cheaper tuition fees will also be a plus. For example, Makerere University in Uganda has a good reputation, as do several universities in South Africa. Studying in South America will give you a good command of Spanish if you don't already have this. The BRAC University in Bangladesh and the Universidad de Los Andes in Columbia are two of 22 Universities worldwide offering the Global Master's in Development Practice (MDP) for example.

I did a Master's in Adult Education in UCT in Cape Town. I was working in Zimbabwe at the time, and wanted to stay in the region. I found the sociology department at UCT very vibrant. The Master's offered several options, including courses on radical pedagogy and informal learning, which is what I was really interested in. The recent history of the country created a very fertile learning ground. Costs were also slightly lower than a similar Master's back in Ireland, although part-time work not so well remunerated.

Sive (Ireland)

If you do chose to study outside your own country, make sure however that the qualifications will be accredited in your own country, if you need them to be. Despite the Bologna process, different countries' qualifications vary, and may not be recognised from one country to another. While this is not a problem for most jobs, it may be for government and UN JPO positions.

My dream was to go to England to complete my Master's, as they are world leaders in the field of development, with a strong academic staff and good resources. I also work with many English people on a daily basis, so understanding their culture and perspectives was also important to me. On the downside it is very expensive as international students are required to pay a lot of money compared to other students and I had to save up for several years to be able to cover tuition fees. At the beginning I didn't think I would be a good candidate for scholarships, and didn't even know where to apply or what the requirements were, such as the International English Language Testing System (IELTS) certificate. After a lot of searching I was very lucky to get a full paid scholarship from the University of Westminster (London). The course that I did and the exposure that I got were an amazing experience for me. I was able to return to my job with new skills such as writing grants, writing good reports and time management. Apply for courses that are relevant to your first degree, occupation and your country's development needs, that way it will be easier to get matching scholarships.

Sam (Uganda)

Course costs: Course tuition fees and living expensive can be prohibitive, or at least plunge students into significant debt. Some scholarships exist and this might be an option to look into. But at the same time there is a huge variation in course costs. Table 6.1 gives some examples of the ranges.

Table 6.1 represents the course costs of full-time courses. But many of these can actually also be done as part-time courses, or online. The course costs will be similar but spread out over a longer period.

I was seeking a programme with a diverse international perspective and hoped to study internationally, outside of my home country (USA). I looked into programmes in Canada, Germany, South Korea, Singapore, Australia. I chose to do my Master's in International Development at Lund University in Sweden.

TABLE 6.1 Comparison of courses and costs, accurate as of 2013 admissions. All prices quoted in USD using September 2013 exchange rates

University	Country	Course	Duration	Tuition cost	Living cost
University of Copenhagen	Denmark	Master of Disaster Management	12 months	$19,700 for EU and Switzerland nationals and $26,300 for others (whereas other MAs, such as African Studies are free for EU students and $9,300 for non-EU students)	$1,000 monthly
Université Cheikh Anta Diop (UCAD)	Senegal	Master's in Development Practice (MDP)	25 months	$12,000 ($6,000 annually)	$500 monthly
School of African and Oriental Sciences (SOAS)	London	MSc in Violence Conflict and Development	12 months	$13,700 for EU students and $23,600 for international students	$1,500 monthly
Wageningen University	Netherlands	Master International Land and Water Management	24 months	$2,350 for EU students and $17,600 for non-EU students	$1,200 monthly
University of New South Wales	Australia	Development Studies	24 months	$490 per credit and 48 credits required for award ($23,520)	$1,000 monthly

The LUMID (Lund University Masters in International Development) programme in Sweden was my first choice for many reasons: it attracted students from all around the world with diverse backgrounds; it was aimed at individuals who already had experience in the field; I was interested in the Swedish perspective of development and the programme had a strong management and practical component. It was a two-year course and a large portion of the second year was spent on an internship and field research in a developing country. Another big plus was that at the time the course was free, which was fantastic (however, now free tuition is only offered for EU students). The downside was that I have received almost no post-programme career support and the support for finding internships was very little. Since graduating in 2011, I have returned to Vietnam, where I lived prior to the LUMID program. I am working as a director of a grassroots organisation.

Annetta (USA)

Intensity: Should you study full time, or part time? On site or distance learning? Studying while you are working can be a good option, as you have the advantage of an income and being able to apply your knowledge to a practical experience. But it also has its challenges: finding the time to dedicate to your studies after a busy week at work and taking full advantage of the opportunities available to you as a student. Some prefer to focus on a full-time Master's to get the most out of the student experience.

An ever-increasing number of courses are now available through distance learning. With a good internet connection you can study through an interactive online portal, enabling you to connect to students all around the globe. Some programmes may require you to meet physically for a few weeks a year, and have to do an examination in an accredited institution, while others may be entirely via coursework.

I am doing an online Master's in Agriculture Economics slowly over five years to fit it around other work and commitments. Even though I haven't finished the Master's I have it on my CV and it looks good to employers.

Debora (Canada)

DID YOU KNOW? MA, MSC, MPHIL: WHAT IS THE DIFFERENCE?

MA stands for Master of Arts and MSc is a Master of Science. Some Development Studies degrees are MAs, others are MScs, depending mostly on the university rather than reflective of content. Both are taught degrees. An MPhil is different; a Master of Philosophy is largely a research focused degree. Some get this through the first year of a PhD, for others it's a course in their own right. The letters that you have after your name don't usually make any difference to employers.

University and course design: Some Master's are very theoretical and provide a good academic grounding, others are more applied. Some things to look out for when selecting the right course for you include:

- *Theoretical/practical?* The study of democratic processes in Africa may be fascinating and useful for a PhD or an analyst role. When it comes to working for an NGO on the ground, knowledge of how to write a good project proposal, develop indicators on a log frame or do an impact evaluation is necessary. More prestigious universities are likely to offer more theoretical courses, while practical courses may come from smaller less well-known institutions. Look for a course that emphasises employability, and has been designed in partnership with employers.

- *Who are the teachers?* Academics or practitioners. Depending on your career aspirations, you may prefer to find a course where practitioners are among the teaching staff, and not just approach development from an academic perspective.

- *Time spent in the field?* As so many jobs require experience, a Master's course that supports internships or overseas placements can be a big plus, although this usually requires some compromises (e.g. a two-year course instead of one year, fewer specialism module choices).

- *Who are the other students?* Number of students on the course and their typical profile is also worth considering to make sure that you are a good fit. You will learn a lot from your fellow students and they will become part of your future network, so the more experience they have the better. You want enough students on the course, but it should be small enough to become intimate and get to know each other.

- *The university's reputation*: Choosing your university carefully is important; some have a good reputation that will follow you in your career. Moreover, you are likely to gain as much from the other students on the course as from the actual course content. So check out who the other students on the course are. Recent graduates or people with several years' experience? From developed or developing countries?

> I wanted to go into Public Health and did a Master's in Health Promotion at the London School of Hygiene and Tropical Medicine (LSHTM). I knew it was one of the best places to do this Master's, and believe that the fact that I studied there has opened many doors for me. One job I applied for wanted a medically trained person, with a Master's in Public Health (MPH) qualification. I wasn't medically trained, but when they saw that I had studied at LSHTM they waived this criterion and I got the job. I would advise you to choose your institution well.
>
> *Alessia (Italy)*

Making the most out of your time as a student

A Master's course provides the ideal platform from which to launch and develop your career so take full use of the opportunities available to you while you are a student. Remember that your time as a student is less about the qualification and more about the doors that it will open. Here are some suggestions that you can adopt.

Gear your dissertation towards your area of interest

Most Master's courses will have a research dissertation or extended assignment. Don't see it just as a tick-box requirement to obtain your qualification, but as an opportunity to gain specialist knowledge in an area, make contacts and even publish your work. If you gear it toward the area of work you would like to work in, it can open up many doors.

For example, if you have an interest in monitoring and evaluation, do a research-based Master's using quantitative and qualitative techniques, which could be a good platform for this type of job and give you something to talk about in the interview. Better still, you could evaluate a project for an organisation, making your research valuable both to them and to you. Find out what the new and emerging trends are, and focus on these areas. Chances are that fewer people will have specialised knowledge of these areas, so when it comes to applying for jobs, you will be at a competitive advantage.

> I specialised in the Kyoto protocol for my Master's dissertation, just after it had come out. It was very timely and as a result I knew much more about it than my senior peers, so it put me in a good position for jobs.
>
> *Courtney (USA)*

Your dissertation is also the ideal time to reach out to and make contact with influential people in the field. If you are able to make a good impression it may stand you in good stead when it comes to job searching, especially if they have anything to do with the recruitment process.

> I was interested in entrepreneurship and private sector development and as part of my dissertation interviewed a few leaders in the field, including the heads of NGOs. I also made sure I sent them a copy of my dissertation, and a summary paper. When I started looking for jobs, it was one of these contacts who forwarded me the vacancy for the role I would eventually get.
>
> *Anonymous*

Complete an internship

Several organisations that offer internships have the eligibility requirements of being a full-time student. You may have to cover your own costs (flight and subsistence) and there is absolutely no guarantee of a job at the end, but it will provide that vital experience, and boost your CV. If you don't already have field experience, internships are the ideal way to make a start on this and will help you with vital contacts, gain insider knowledge about an organisation and help inform your future career decisions. If you already have some professional experience, going back to being an intern may be demoralising, but chose an organisation that takes interns seriously so you will have the opportunity to engage with some meaningful work.

> In the third semester of the NOHA programme, you are encouraged to do a three-month internship while you write your thesis (the alternative is to study further in a partner university). I interned at a UN agency for six months (extending beyond the initial three months). I worked as part of a small team that concentrated on the training and partnership aspects of a section's work. Following the internship, I was employed as a contractor, doing broadly similar work, though I was more involved in training and could now travel to the actual courses (UN interns cannot travel officially).
>
> *Anonymous (UK)*

Attend conferences and events

Instead of just sticking your nose in a book, trying to keep up to speed with assignments, attend relevant conferences and events and use them as networking opportunities, as well as a way to understand the latest areas of research. It's a great way to meet people, find emerging themes to focus your dissertation on, which eventually might lead on to employment opportunities. Often heavily discounted rates are available to students and this is a great way to meet people and position your research in the most up-to-date narratives.

Make full use of university services and events

Universities often have career evenings, alumni who return to do talks, practitioners who come in and give advice, etc. Career Services are also the ideal situation to get expert support and advice for planning internships and networks. Attend recruiter talks and events.

> I did a Master's in Public Administration (MPA) in Non-profit Management and Policy at New York University (NYU). I applied to NYU because I was particularly interested in social entrepreneurship, and the university then, along with Harvard, was the only institution offering a postgraduate fellowship in the field. Education in the US is expensive, but there are some scholarships available. I was fortunate to be awarded one of these.
>
> I found that there was a definite difference between UK and US teaching styles – there was a greater emphasis on practical learning through guest

speakers who shared their professional experience, group learning and interactive lectures. There were a lot of opportunities for networking and career development, possibly because American universities are ranked according to graduate employment rates. At NYU there were daily talks by recruiters as well as graduates, and the careers service is very proactive in guiding students to make successful applications, as well as finding internships.

The majority of my fellow students have gone on to work in non-profit or public sector organisations, as ultimately our programme provided us with the tools and skills specifically to work in those fields. There are some, like myself, that are now leading smaller organisations, and others that have gone into mid-level management in larger organisations with a range of organisations, many of which focus on international development, such as BRAC, SPARK.

David (UK)

Network

One of the biggest assets of a Master's will be your peers, who come to the course from a wide background of skills and experience. Learn and connect with them as much as possible, as they will provide part of your future network.

And don't neglect your tutors, many of whom will be well connected to the world outside your university; if they see you as a bright and eager student, they will probably be more than happy to recommend you to some of their contacts, for internships, short-term assignments or even jobs.

When I first received information about my course mates – especially around the amount of experience they had – I knew that the taught basics would be greatly enhanced by informal interactions.

Anonymous (UK)

Some thoughts on PhDs

Given how many people are now getting Master's to stand out from those with Bachelor's, will the PhD become the new Master's? After your Master's course, you may feel a PhD is a natural step. As you think about advancing in your career, you may want to add a PhD to your portfolio of qualifications. But outside of academic circles, will doing a PhD be an advantage? Will it enhance your chances of employability?

It will depend on your career aspirations. Some employers (such as the World Bank) place a greater emphasis on qualifications than others. Many developing countries also value paper qualifications highly, so in many cases your status will rise with PhD letters behind your name. If you want to go into an advisory role, or a high-level consultancy position, doing a PhD can also be an advantage – and is sometimes required.

My experience over the last decade is that in developing countries, many senior officials have a doctoral qualification – often thanks to generous scholarships – at good overseas universities. So as an advisor to such, it is best to have a doctoral qualification. In one of my consultancies, the beneficiary specified that all consultants had to have a doctoral qualification.

Geoff (Australia)

For those who want to work in the field (and not in academia) it's advisable not to do a PhD straight after an undergraduate degree and a Master's. With a purely academic background, little hands-on experience and several years in education you may be viewed as a bookworm by employers, good with theory but less so with practice. You will probably have to go into entry level positions, or employers may be discouraged from recruiting you as you may command a higher salary, while performing the same tasks as a Master's graduate. Get experience first, find an area that you are truly passionate about and pursue it as a PhD later on.

A PhD may seem daunting to many. Spending three or four years of your life focusing on just one area. But it needn't be so scary. In fact it is the greatest luxury you will ever have: to have all that time to spend thinking about just one area in such detail. You might be worried about your boredom threshold but then, you can spend a whole day in bed reading and be called productive for doing do! What else allows for that? If you want to go into a research job, I would definitely recommend investing your time in doing a PhD since a Master's will not give you the research skills you will require for the role. And after you have a PhD the jump is quite significant in terms of the level of jobs you can access.

Shandana (Fellow, IDS)

But once you have started doing exciting technical work, it can be very hard to take the time off to pursue a doctorate. It is possible (and advantageous) to combine both, especially with a very practical doctorate degree. This can lead to a worthwhile thesis that is based on field work within your current area of work. The result is likely to be a very timely and relevant piece of work that is beneficial both to you, your academic institution and the field of research.

After ten years of working in international development education I am now pursuing a part-time doctorate in Global Education. I am combining work in the field as a Senior Education Advisor on a donor-funded project while taking online and on-site courses at the University of Southern California and Hong Kong University. My dissertation will be closely aligned with the field project and policy reforms. I'm hoping this will provide the best of both worlds, although combining work with dissertation writing will be a challenge. A doctorate is not necessary to be successful in the field (I would suggest at least a Master's with a focus on impact evaluation methods) but as an educationalist, we all thirst for more knowledge.

Brenda (USA)

Final thoughts

There are many ways to advance in your career and be a better development or humanitarian professional without gaining a Master's course or a PhD. Continuous professional development is, however, important and should be something you prioritise, but there are many different ways to do this: keeping abreast of recent developments, liaising with colleagues, reading, self-directed learning, short courses, participating in working groups, doing independent research and presenting findings and so on. Academic qualifications are only one small part of it, and Chapter 9, Advancing in your career, will give you some other ideas.

PLATE 4 Children in rural Uganda carrying water from the river to their homes. In much of Africa carrying water is a daily task for children and can limit their access to schooling as well as exposing them to hazards. These children carry containers weighing close to their own body weight for around 3km each day.

Source: Phil Crosby

7

VOLUNTEERING AND INTERNSHIPS

People come to volunteering for a whole host of reasons. For some it's a way to gain experience within the sector and overcome the no-experience-no-job catch-22. Volunteering with the right organisation can be a good way to build your CV and gain the contacts, experience and insight necessary to propel you along with your career. If you are already mid career, and looking to make a side step into the field, it can help with this transition and can also be an opportunity for you to gain access to the sector, understand whether it's a career you really want to pursue and what area to specialise in. For others, usually mid or end career, volunteering may simply be a chance to share your skills and experience for the benefit of others while experiencing a new way of life.

If you are considering doing some volunteer work – be it at the start, middle or end of your career – or an internship (covered in the second part) – this chapter is for you. It will help you to understand the different types of opportunities out there, how to choose one that is right for you, and finally, once you have got it, how to make the most out of it.

Types of volunteer opportunities

People often think of volunteering as working without financial remuneration. For many positions this is true but there are large discrepancies. Some organisations may ask you to cover all your own costs, or certain expenses (travel, lunch allowance) may be reimbursed. Others may ask you to pay a substantial fee for the experience, while others will cover all your costs and give you a local wage or generous living stipend. It all depends on what you have to offer, the organisation you volunteer through and the duration. This section looks at some of the different types of volunteer opportunities available.

Volunteer . . . in your spare time

Any free time that you have while studying or working can be put to good use by volunteering. If you are serious about a transition to a career in the field, this is highly recommended, as there are few opportunity costs and it can give your career a boost, often in indirect ways. If you are lucky enough to live in a place where there are a lot of international organisations, identify the ones that work in areas you directly want to gain experience in and target them first. If it's not exactly what you want to do, don't worry too much, you never know what it might lead on to. There are also an increasing number of volunteer opportunities online – for example for doing research into particular topics.

> I went on from an undergraduate course straight to an MA (in Conflict, Security and Development at King's College, London). After I graduated I became despondent about looking for jobs. I didn't have the experience necessary to get even the most basic of jobs. It seemed that my first hurdle would be to get an internship, but these are also very competitive. I then applied for some volunteer work with a small human rights organisation based in London, called Waging Peace. I heard about them through a friend, who had heard the founder speak. I started working on a China in Africa report, coordinating other volunteers and editing/writing parts of the report (alongside doing 40 hours a week in a clothes shop). It is about always being available and working hard when something does come your way, personal relationships and getting on well with the person you are working with. And luck – being in the right place at the right time. When a staff member was leaving the organisation, I was offered the opportunity to take over her role.
>
> *Sophie (UK)*

Most organisations will advertise for volunteers when they have specific pieces of work that come up or they need additional help, or they might run regular recruitment drives. These will be advertised on the organisation's own website or job advert sites – such as CharityJOB, Devex, Idealist, see Chapter 8 – search under volunteer posts.

You can also use your networks to find a volunteer placement, or target specific organisations: send a CV and ask whether there are any volunteer opportunities or pieces of work you can help them with. Ad hoc volunteer opportunities of this type are usually easier to find with the more flexible smaller organisations, who can be short staffed and would benefit from a capable offer.

You pay . . . to volunteer

Several organisations will arrange an international placement for you in exchange for a fee. They may also provide support on the ground and help with travel and accommodation arrangements. If you are new to international work and want a guided experience, volunteering through one of these organisations will make the planning much easier. The duration may be from as little as two weeks to six months or longer.

Be aware that you will be the one benefiting the most from this type of work – a legitimate reason to pay for this experience – as inexperienced volunteers require significant resources to manage. But before paying for a *volontourism* experience do ask several questions, including:

- *Where does the money go?* Volunteering has become a multimillion dollar industry – making money out of poverty. Some organisations operate as commercial *gap-year* style companies while others have community development as their main remit and use volunteers to enhance their work, charging only for the real costs of arranging and managing a placement.
- *What support is offered?* Comprehensive preparation for a field placement is important to make the most of it, as is network of supervision on the ground and support in case of emergency.
- *What will I be doing?* Watch out for instances where you may be doing more harm than good. Ask yourself (and the organisations) if and how you will be adding value, rather than taking jobs away from local workers. Short volunteer stints at orphanages also come with a warning as they can create emotional attachment issues with the children, with real consequences on these lives. Even some medical missions, if poorly organised, can impact negatively on local systems and health care providers.

If you decide to go down this route, make sure you go with a reputable organisation, compare costs with what they offer, understand how the placements are organised and have a clear understanding of your role before you go. Some examples of development oriented organisations include:

- Restless Development – a youth-led development organisation, www.restless development.org/. They recruit international and national volunteers to work in the 11 countries they operate from. Their International Citizen Service (ICS) is funded by DFID and is open to 18–25 year olds. Volunteers must be either a British citizen, a European Economic Area (EEA) citizen residing in the UK or have indefinite leave to remain in the UK. Most volunteers are asked to raise £800.
- 2waydevelopment – based in the UK, but accepting volunteers from all around the world. They charge a flat fee of £850 for arranging a placement (3–24 months) plus your living costs. It covers pre- and post-departure support; they will offer you three handpicked volunteer opportunities to choose from and CV and job advice once you return. Minimum age is 21.
- Global Volunteers – www.globalvolunteers.org. If you are after a volunteer vacation between two and six weeks, you can try an organisation like this. US based and founded in 1984.

In the year between my undergraduate degree and my Master's (in Develop-ment and Trade) I volunteered. I decided to do so with Restless Development. I did have to pay for the placement but it was worth it. It was very well organised, I got a lot of support and got exposure to a wide variety of different work. It was an eight-month placement in Nepal, including a one-month introductory training course that included learning key development techniques, such as participatory rural appraisal (PRA), cultural sensitivity and language training. I was paired with two Nepali girls my age, and we were partnered with local community-based and civil society organizations. Our focus was community awareness raising campaigns (dramas, rallies, workshops, video days etc.) on issues relating to nutrition, child trafficking and sexual health. I think when you have relatively little experience to offer, and want a well-planned and supported placement, it is worth paying for this.

Catharine (UK)

Short-term, expenses-covered volunteering

Organisations that recruit for short-term (two weeks to six months or more) volunteer opportunities where your expenses are paid tend to be reserved for those with substantial professional experience behind them, such as clinicians, engineers and so on, ideally complemented with some developing-country exposure. If you are in this category and want to make the transition to full-time work in this sector, these short-term posts can give you a snapshot of what the work might be like and give you some experience. Or they can be usefully combined with full-time work back in your home country.

You could first turn to your professional *sans frontières/without borders* – if they exist. From *architects without borders*, *bankers without borders* (linked to the Grameen Bank microfinance organisation) to *vets beyond borders*, your profession may have a chapter. Some of these receive donor funds for their project to cover costs of their volunteers. You can also look for organisations that specialise in your area and see whether they recruit volunteers.

Humanitarian organisations often recruit short-term volunteers when an emergency strikes. But here your professional skills will not be enough; you will have to have some experience of working in an emergency context and training in it. RedR's 'So you think you want to be a relief worker' one-day course might help you to understand the sector a bit more.

Longer-term paid volunteering positions

There are a number of development organisations that recruit long-term volunteers to support their programmes and partners for six months to two years or longer. This is a substantial investment of time, but long-term placements have several advantages both for you and for the organisation you are working with: it will allow you to gain a deep understanding of the community you are working with, the local context and the development challenges. You will also be able to build up substantial skills and take on significant responsibilities, which is why employers look favourably on these types of roles. You'll build strong friendships and maybe even learn the language. Evaluations consistently show that the local organisations also benefit more from long-term volunteers.

You will need to have transferable skills for these roles, and the more experienced you are the better. Organisations will often provide you with a living stipend (usually the equivalent of a local salary), medical coverage, flights and pre-departure training.

Some of the larger organisations recruiting for long-term volunteer placements are listed below. But note that this is still no easy route, with significant competition for jobs. If you are successful, the application to placement time can be over 12 months, so plan early. Check out the main volunteering organisation from your country, which may include organisations such as:

- Voluntary Services Overseas (VSO) is an international development organisation that tackles poverty through volunteering. It mobilises skilled professionals to work with partner organisations in over 40 countries. Placements tend to last 1–2 years although a large number of volunteers extend for longer. Competition is fierce as VSO receives over 10,000 applications a year for around 500 funded long-term overseas placements. No nationality restriction for volunteers. Apply online at www.vso.org (rolling recruitment).
- United National Volunteers (UNV) work towards global peace and development. They recruit professionals from a broad range of specialisations and in 2011 UNV engaged 7,303 UN volunteers on a total of 7,708 assignments. With an average age of 38 years and the requisite professional skills and qualifications of some 5 to 10 years' relevant experience, UN volunteers served in 132 countries in 2011 and came themselves from 162. The majority – 81 per cent – come from developing countries themselves and around 30 per cent volunteer in their own country. No nationality restrictions for volunteers. Placement usually lasts up to two years. www.unvolunteers.org
- For US nationals, the Peace Corps Volunteers is a very popular and highly regarded scheme. Volunteers work to address changing and complex needs in education, health, HIV and AIDS, business, information technology, agriculture and the environment. One of the goals of the Peace Corps is to help the people of other countries gain a better understanding of Americans and their multicultural society and is thus only open to American Citizens. They provide excellent language training as volunteers tend to live in remote communities. Apply by the end of September to start a placement the following year. Returned volunteers and other professionals can also take part in the Peace Corps Response – challenging, short-term (3–12 month) assignments in various programme areas around the world. www.peacecorps.gov
- Australian Volunteers for International Development – open to Australian and New Zealand nationals and residents. www.volunteering.austraining.com. au/faqs – skilled volunteer placements. Apply for a specific post.

- World Friends Korea, inaugurated in 2009, is the Korean equivalent of Peace Corps and has priorities in education, health, IT and community development, with placements in Asia, Africa and Latin America.
- PUM Netherlands senior experts www.pum.nl connects entrepreneurs in developing countries and emerging markets with senior experts from the Netherlands, each of whom has gained at least 30 years of experience in a business environment. These senior experts voluntarily devote their knowledge to the execution of short-term, solid consultancy projects.

> I had the good fortune to have a rewarding career that allowed me to retire at 51. Given my low boredom threshold and my lifelong volunteer efforts I knew I'd need to find meaningful 'work' for my retirement years. Peace Corps had always been in the back of my mind as an option.
>
> I'm a passionate advocate that everyone who has the opportunity to go abroad for 27 months (without obligations holding them back) should pursue the Peace Corps. Why is that?
>
> For my age bracket it's the ultimate anti-aging elixir. You'll be surrounded by younger volunteers whose energy will invigorate you. Second, you have a lifetime of experiences behind you to share. Third, it's a terrific brain teaser. You'll find yourself in situations you never could imagine and have to work your way through them . . . in another language!
>
> I was assigned to the small village of Ribat El Kheir, about one hour from Fez, Morocco. The community opened its arms to me, generously supported me personally as well as through my projects and I left a piece of my heart behind. The good news is that the women I worked with (in a weaving cooperative) are continuing the initiatives we started together and have increased their access to market for their products. Ultimately it's not what you do while you're there, it's what they do when you're gone.
>
> Many of my fellow volunteers were recent graduates, interested in pursuing a career in international development and using Peace Corps as a stepping stone to competitive paid positions. One that comes to mind subsequently landed a great job with the MCC.
>
> *Lynn (USA)*

Do-it-yourself volunteering abroad

Combining work and volunteering with an organisation back at home may pay off, but you will miss out on the international experience – a dimension needed for many jobs. If you want to go overseas without the constraints of a *matching* organisation, it is possible to arrange your own volunteer placement overseas.

It can be a challenge to arrange such work from the other side of the world, with little knowledge of which organisations are on the ground. Once you have some focus (country and area of interest), start using contacts as not all local organisations will have websites (or websites that rank highly in searches). If you don't have the contacts, make some. Approach organisations back at home who work with your particular country of interest and find out about their partners; university professors who have carried out research there; or see if you can find someone useful to guide you at your embassy in that country.

It is possible – but a greater gamble – to arrange a placement once you are in the country. This can be a good strategy if you are combining it with travel, but it is advisable to try to line something up before you go. Check out visa regulations as well; for some countries you will need a volunteer visa, and in some cases will need to have an organisation sponsor you for this before you travel.

> The advice I kept getting from people when I was looking for a job was that I need international experience. I would read articles by aid workers saying you need to 'just go to the country'. How could I do that? With no connections, no job, without much money behind me, I couldn't just get up and go! After finishing my Master's at Edinburgh University on the Anthropology of Health and Illness, I struggled with internships that weren't really getting me where I wanted to go.
>
> Eventually I decided to just go for it. I made contact with a small health NGO in Rwanda through a friend, and arranged a one-month volunteer placement. My intention was to stay and try and look for opportunities while I was there, hopefully leading to longer-term volunteer or paid employment. Once in Rwanda I decided to keep working with the local NGO.
>
> Three years later, I am still in Rwanda, working as a consultant for the UN. How did that happen? During my placement with the local NGO, I got involved in writing some grant applications for them relating to HIV. One of the grants was chosen and after that I got a call from the UN asking me whether I would be interested in doing some consultancy work for them.
>
> I did a few short-term consultancies for the UN, including writing a strategic plan on gender and HIV for the Rwandan government. They didn't have many people specialising in gender in the country available and who knew the situation on the ground well. One thing led to another and then I was offered a longer-term consultancy with the UN.
>
> This strategy can't work in all countries. Rwanda is small – at the time it was relatively easy to get a work permit as a foreigner and everyone knows everyone else, which makes networking very easy. I have a friend working in Laos who had a similar experience.
>
> *Kate (USA)*

The advantage of this technique is that you have complete say over the country you go to and type of organisations you volunteer for and the duration. But there will probably be less structure to the volunteer post and you will have to make it into what you want it to be, so being able to offer the organisation particular expertise will be essential. Don't expect people to drop their jobs to show you yours – you will have to be able to work quite independently and have a lot of initiative.

FIND OUT MORE! RESOURCES FOR VOLUNTEERS

There are several books and resources out there for those wanting to volunteer for development organisations. You could try looking at some of these:

- Fabio Ausender (2008) *World Volunteers: The World Guide to Humanitarian and Development Volunteering*, 4th edn. Crimson Publishing;
- Charlotte Hindle (2007) *Volunteer: A Traveller's Guide to Making a Difference Around the World* (Lonely Planet General Reference);
- *Volunteering for Development*, produced by the UK World Service Enquiry, includes a list of over 350 organisations that send people overseas. ISBN: 978 0 9558393 1 3. Available only from their website on www.wse.org.uk/guides/vol.htm.

Internships

There can be a fine line between internships and volunteer placements, although in general internships are geared towards providing current or recent graduates with supervised practical training, and are even a requirement for some Master's courses. They tend to be full time for a fixed period and *some* may offer a living stipend or basic salary. Certain organisations require you to be enrolled in a full-time study programme to be eligible – others target recent graduates – and there is often an age restriction.

Internships – especially the paid ones – are often highly competitive, but if you are lucky enough to be selected they can provide a fantastic way to step inside an organisation that others are not able to gain access to.

UN and other multilateral organisations

Apart from the UN Volunteers scheme (see above) most of the UN offices offer internship schemes. The United National Headquarters for example offer internships at all their major locations (New York, Beirut, Addis Ababa, Santiago . . .). These are ideal if you are thinking of entering the world of diplomacy and public policy and want to experience the day-to-day working of the UN HQ. A common requirement is that you be enrolled in a Master's or a PhD programme, and are able to cover your own costs as they are not paid. Internships can be between two and six months. You can apply by visiting the United Nations Career Portal. https://careers.un.org/. Other United Nations funds and programmes also offer internships (such as the UNDP, UNFPA, UNICEF, WHO, WFP . . .) so check out each of their websites for specific details.

The World Bank offers paid summer and winter internships in their Washington office with potential duty stations. The minimum duration is four weeks and students of particular specialities are sought. Again, candidates must be enrolled in a programme and be returning to full-time studying after their internship. Similarly, most of the other multilateral organisations offer internships.

Some of these organisations also offer highly competitive Junior Professional programmes usually aimed at people under 32 years of age with a relevant speciality and experience. Check out each organisation's website for details, criteria and application timescales.

> After my Master's I did an internship with the World Food Programme (WFP) in Zambia. This was through the 'Global Experience Programme', a partnership between TNT and the UN emergency relief organisation. I was supposed to work on logistics (distributing food aid to refugee camps and schools) but soon I started to realise that the bleak reality of working in a country office far removed from the actual conflict means you are entering numbers in a system tracking the movement of food trucks, staring at Excel sheets all day.
>
> A new ground-breaking programme just started, called Purchase for Progress (P4P), and I asked whether I could join in setting up the programme. P4P was a pilot aimed at changing the way WFP buys its food. Instead of shipping everything from the US, the programme sought innovative ways to buy everything locally from small-scale farmers, boosting their income and linking them to the market.
>
> This turned out to be a fascinating experience that introduced me to the world of agriculture and markets. After that I worked as a consultant for the newly established agricultural market commodity exchange in Zambia, and subsequently worked for the P4P programme in Rwanda. That experience really shaped my career in agriculture and value chains.
>
> *Janno (Netherlands)*

Government-sponsored programmes

Recognising the growing interest among young people in the development and humanitarian sectors – and the challenges they face when entering the job market – many governments have started supporting internships or training schemes. The aim of these is to equip recent graduates with specific skills to enhance their employability in the sector. In some cases only nationals of that country will be eligible, although within Europe, all European and Swiss residents and nationals can apply.

The following examples illustrate some of the schemes available; check out what is available within your own country.

- *Canada* – The International Youth Internship Programme (IYIP) is part of the Government of Canada's Youth Employment Strategy. Through this programme, Canadian NGOs are provided with funding to hire young Canadians, aged 19–30, as interns for a period of 6 to 12 months. In turn, these Canadian-based NGOs match the intern up with a partner organisation that is operating aboard.

- *Germany* – the German Technical Cooperation Agency (GIZ) has many programmes. They offer one to six month internships in departments across the company, both in Germany and abroad. Internships are open to undergraduate and postgraduate students. They also run a Development Cooperation Trainee Programme, which lasts for 17 months, is paid, and is open to people of any nationality with a strong command of spoken and written German.
- *Japan* – JICA offers an internship programme. Applicants for the programme require knowledge of the Japanese language to apply.
- *United Kingdom* – the DFID Graduate Development Scheme was launched in January 2012 and offers a paid, 50-week UK-based programme offering a diverse range of opportunities in areas from humanitarian response to private sector development. Its aim is to enhance employability in the sector and no job is guaranteed with DFID at the end. The scheme is open to UK, EEA and Swiss nationals as well as Commonwealth citizens with indefinite right to remain in the UK without employment restrictions.

In February 2012 I started a six-month internship in Arusha, Tanzania as a part of the IYIP programme [see above]. I had previously completed a BA in Political Science.

I interned with a Tanzanian NGO that provides services to street involved children. As an intern, I was exposed to the daily operations, rewards, challenges and frustrations associated with development work. Through my placement I developed technical skills that are an integral part of NGO work, such as monitoring and evaluation techniques, and proposal writing. My exposure to an unfamiliar and challenging work environment was invaluable as it helped me determine whether this type of work is for me. As in many NGOs, my organisation had limited financial and human resources, constraining its efforts to provide services to its target group. Its limited capacity also hindered its staff's ability to assist me in my assigned project, bringing about personal frustrations. Nevertheless, this lack of support also taught me that technical expertise, not merely a desire to help out, is paramount, thereby validating the ultimate goal of any internship: to gain experience and skills. Overall, I would recommend an overseas internship placement to anyone who wants to see whether development work is right for them; just make sure your expectations are realistic, as the biggest change will happen within yourself.

Now that I have completed my placement, I intend to study Disaster Management. The real-life lessons I learned in Tanzania will undoubtedly prove useful in both the classroom and the field.

Fraser (Canada)

Other organisations listed in Chapter 4 (Who could you work for?) including NGOs, development consultancies, think tanks and trusts and foundations may also offer internship opportunities. Some have structured schemes with regular recruitment drives, while others might offer internships as and when they are needed. Some are paid and others are voluntary. Opportunities are usually advertised on an organisation's own pages and job search sites (see Chapter 8). Most of these give you the option of searching for volunteer positions and internships only. Terminology is often blurry and if a role is called internship it is not necessarily better than one termed volunteer, so read the description. You can also research and make a list of the organisations you are interested in (as well as countries or sectors) and search for volunteer positions in these specific areas. Use your contacts as well.

> As an aspiring development worker it took me quite some time to acknowledge that I was often the one benefiting the most from my volunteering (consider the ethics around an international volunteer building a well, for example). But one area I do feel we – as Western volunteers – can support is in building the capacity (or training) of local staff. Regardless of our technical knowledge, chances are that we can support skills such as computer literacy, working with Excel, general time management, efficiency and working to deadlines, etc. In my case, I felt that one of the most useful legacies I left behind was training the local NGO staff. The work ethics I showed a small office in Kathmandu made a lasting impression on staff. I also spent a lot of time teaching the programme officer formatting in Word and Excel which will have a lasting impact and also give them a future advantage in the local job market.
>
> *Olivia (UK)*

Assessing opportunities

Deciding to apply for, and accept, a volunteer position or internship requires careful research and deliberation. Be clear what you want out of the experience and whether a particular organisation and position will offer this. After all, you can't spend all your life doing full-time unpaid work and there are only so many summer internships you can do while you are a student, so there is an opportunity cost to each position you accept. At the end of this chapter is a worksheet (Figure 7.1) which will help you to plan your time as a volunteer. Complete section 1 (your motivations for volunteering) and section 2 (your ideal volunteer role) before reading on.

> Volunteering doesn't always feel good and can be demeaning as you are not taken seriously. At the age of 32, with three years' experience of working in Africa and eight years in social work, I returned to the US to find a job in development. I decided to volunteer one day a week with an organisation I had had three unsuccessful job interviews with, as it might be a foot in the

door. The volunteer programme wasn't very coordinated and every day we were about ten volunteers going around asking everyone whether they needed help. You really have to decide whether the pros out-weigh the cons. For me they did as I was able to meet people and viewed it as a networking oppor-tunity, even though I did have to swallow my pride when being bossed around by 24-year-olds, who don't know who you are or what experience you have. It was tough. I would advise looking for a volunteer programme that is well set up and where you are assigned to a particular department, so that you can get more engaged in the work and don't find yourself with tasks such as filling up the coffee machine.

Jessi (USA)

You should approach finding a volunteer opportunity or an internship much as you would a job. Overseas positions and ones that offer a living stipend are often as competitive as paid positions so you will have to invest substantial time in the application process.

Table 7.1 highlights a few of the reasons why you might be doing this type of work, and some of the considerations to take into account for each. Now you have thought about your own motivations, compare them to those in Table 7.1 and what you might need to think about when looking for roles.

Before you accept a specific role, check that it matches your criteria and addresses the reason why you wanted to volunteer in the first place (what you wrote in section 1 of the volunteering worksheet). Ask more questions about the role, what you will be doing, what your own learning outcomes might be and the level of supervision. If the position you have been offered doesn't seem to be able to offer you what you are looking for, keep on looking. There are volunteering and internship programmes to suit everyone.

CAREER CLINIC! WILL VOLUNTEERING HELP ME TO ENTER THE JOB MARKET?

Some feel that the culture of volunteering and internships as a route into the development sector is exploitative and unegalitarian (as only those with means have the option to work for little or no pay). It's important to weigh up the benefits of a volunteer role over other experiences. If your main motivation for volunteering is to transition into a job in development, you need to analyse your skills and the position carefully to decide whether its right for you. Refer to the T-shaped person sought after by HR managers in Chapter 9. If the volunteer role complements your other skills, it may well be the break you need to appear on the job interview shortlist. Carlos, for example, was an architect with several years of experience in Spain and England behind him, who wanted to make the transition into international development work.

He volunteered on a housing project in Asia, in addition to doing a Master's. This experience enabled him to make the transition.

But if you don't have many skills or experience to offer at this stage, volunteering alone is unlikely to lead on to anything just yet – unless you just happen to be in the right place at the right time. It can also help you to gain clarity about what you want to do and the direction you want to go in by exposing you to different types of jobs in the sectors, for example. Zack for example, who interned at the International Child Protection Organization in London for six months while doing an MSc in Violence Conflict and Development at the School of African and Oriental Sciences (SOAS) said:

> It was evident the internship wasn't going to lead on to any kind of job, but it did make me interested in child protection issues. I wished I had been exposed to this type of work before to be able to specialise in this area during my Master's.

Making the most of your time with the organisation

Once you have accepted a position, it is up to you to get the most out of it. It is important to have the right preparation, attitude and mindset to do this.

Before you start

Preparation is important:

- If you haven't yet got it, ask for a job description detailing your activities, tasks and responsibilities. This will allow more thorough preparation. Many organisations may not be able to provide you with an accurate job description, preferring to discuss this with you. If this is the case, read as much as you can about the organisation and their work before so that after a while you can discuss with the organisation what you can offer, matched with the experiences you want to gain out of your placement and what the organisation needs. Write a job description for yourself once in post and agree it with your manager.
- Now you have a better idea about what you will be doing, turn to the volunteer worksheet (Figure 7.1) at the end of this chapter and complete your personal goals: what you hope to get from the experience. Review this halfway through your placement. If there are certain things you aren't achieving, it's a good opportunity to speak to your supervisor about it.
- If you're travelling overseas, this is a perfect time to start networking. Do any friends of friends work in the same town/country? It's always good to have a network of contacts in case things go wrong but at the same time they might

TABLE 7.1 Motivations for doing volunteer work/internships and other considerations

Motivation for doing volunteer work/internships	Think about	Look out for positions/organisations . . .
Share my skills	What skills do you have? Think about your academic qualification and professional skills, but also beyond these: do you have writing, mentoring, motivational skills?	• Where you'll have the opportunity to use your skills • That will allow you to pass your technical skills on to others, building capacity through training or on the job support
Gain a general insight into the world of development or the workings of a particular organisation	What are you most interested in? A birds-eye view, perhaps focusing on the policy side, or working with a grass roots organisation directly with the beneficiaries?	• That work at your level of interest, be it in multilateral or bilateral organisations, INGOs or local community-based organisations (see Chapter 4, Where could you work?) • That work in partnership with other organisations, thus giving you a broader exposure
Get country specific/regional experience	Why are you particularly interested in this country or region? Do you want to specialise in this area? If the region/country is a favourite for development aid chances are there will a lot of job opportunities	• Where you might be involved in analysing policy papers to gain an insight into the political context and local communities to gain a broad overview • Where there might be opportunities to learn the local language
Understand whether I want to specialise within a particular area of development	Which specific areas? Is it a broad area or niche? Is it an area that is in demand for future jobs?	• Where the work specialises in the areas that you are interested in • That will expose you to professionals and mentors working in these areas

Build up particular skills	Which skills? Anything that will enhance your CV is valid – from proposal writing, project management, advocacy, fundraising or utilising your technical skills in a development or humanitarian context	• Where the role description particularly mentions these tasks • Where you will have an experienced supervisor who you can learn from
Complement my academic studies	Are there any opportunities to focus your Master's dissertation on a particular organisation, volunteering for them at the same time?	• That could benefit from your work. For example you might offer to evaluate a particular project, thus gaining experience in evaluation, the organisation's area of focus and obtaining your dissertation at the same time
Network and gain contacts within my area of interest	Are there any experts within the field you would particularly like to work with?	• Where you will have opportunities to liaise with experts within your area of interest, be it at a conference, within the organisations or coordinating partnerships within various organisations • Where you will be able to raise your own profile, through writing reports or news evaluation, and get your name around
Cultural tourism or *volontourism*	In some volunteer posts you may be painting a dormitory or digging a well. Nothing that a local person couldn't do, but a fun time for you. Are you prepared to pay for this experience?	• Where all the arrangements are taken care of for you so that everything runs smoothly and you can enjoy your holiday

also be able to point you towards other opportunities in future. Most people will be all too happy to help you settle in and introduce you to the place. And of course, remember to sort out visas, immunisations, malaria prophylaxis and health insurance and accommodation, ideally before you go!

While you are there

Once your placement is under way, there are certain things that you can do to maximise your experience there:

- Make yourself useful, reliable and willing to help out and get involved in anything, even if it wasn't on your job description. If this involves a day of photocopying, then so be it.
- Your direct manager may already be very busy, and may not have as much time as they would like for training, coordinating an induction and supporting you. Be sensitive to this and use any spare time you have not to browse Facebook, but ask whether there is anything *you* can help out with, or read up on some documents or learn about the issues you are working on.
- Be alert to other things going on in the organisation and if something looks interesting and you have the time, see whether there is a way to get involved. At this stage everything is a great learning opportunity so explore as much as possible.
- Do whatever task you have in hand well. If you prove yourself as competent and easy to work with, people will probably think of you when a new task comes up.
- Seize opportunities when they present themselves and don't be afraid that you don't have enough experience. If someone is needed to write a manual or develop a training course, be bold and step forward. Just ask for help from colleagues to overcome any hurdles.
- Don't get disheartened. Occasionally colleagues might not take you seriously because you are a volunteer. This happened to one doctor in Malawi – colleagues thought she was volunteering because she was a bad doctor and unable to find work back at home. Bridge cultural differences through dialogue.
- If you are in a new country, you might find settling in a little hard. It's important to establish a good network of friends, both local and international. Leisure time and a good work–life balance will increase your productivity at work.

I volunteered at a small street children's centre in East Asia. An opportunity came up to write a manual for activities for early childhood development. I didn't have any formal training in this area, but had a lot of experience working in day care centres during my university years, so I jumped at the opportunity. It was a lot of extra work, and being a volunteer this was of course unpaid. But the manual was very well received and meant my name was spread around.

Early childhood development is an area gaining a lot of popularity, and several international organisations were starting programmes in this area. The manual got picked up by some large organisations and enabled me to network more actively with them. This eventually led on to a paid position in the field.

Carmen (Australia)

And finally, towards the end of your time with the organisation:

- Approach your managers who have been impressed with your work and ask them if they might be willing to act as a referee for future job applications.
- Wrap up any loose ends and write up handover notes if someone is going to be taking over from your work.
- Contact those you have been working closely with during your internship, spread your CV around and let them know you will be looking for a job soon. If you are lucky something may come up.

WORKSHEET: Chapter 7. Volunteering and internships

1. .... 2. ....
....
....
....

1) My main motivations are ...

3. .... 4. ....
....
....
....

2) What would my ideal placement look like?

ORGANISATION

DURATION

INTENSITY

LOCATION

RESPONSIBILITIES

SKILLS EXPERIENCES I WANT TO GAIN

SUPPORT RECEIVED

COSTS

3) My goals and learning objectives for my position are: (try to make these as SMART as possible)

Review whether you have achieved goals halfway through post

1. ....
....

☐ Achieved   ☐ Partially achieved   ☐ Not achieved
ACTIONS:

2. ....
....

☐ Achieved   ☐ Partially achieved   ☐ Not achieved
ACTIONS:

3. ....
....

☐ Achieved   ☐ Partially achieved   ☐ Not achieved
ACTIONS:

4. ....
....

☐ Achieved   ☐ Partially achieved   ☐ Not achieved
ACTIONS:

5. ....
....

☐ Achieved   ☐ Partially achieved   ☐ Not achieved
ACTIONS:

....

SMART: Specific, Measurable, Achievable, Realistic, Timebound

FIGURE 7.1 Worksheet: volunteering and internships.

8

THE JOB SEARCH

Finding – and securing – your ideal job can be an intense and stressful time but this chapter is here to make the process all a little easier. There is no magic bullet to landing a job in the field but this chapter considers some of the different strategies at your disposal. Your CV and covering letter will be your first approach to potential employers, so it's imperative you get them right. The second half of the chapter aims to help you think about how you can improve your application, focusing on the particularities of the sectors. The final section will help you to troubleshoot if your job search is not progressing well.

Job search strategies

The job search can be a gruelling process, time consuming and often dispiriting. When you speak to people now working in the field you will find that many went through the same thing, and persistence paid off for them. The girl who applied for a UN JPO position, got through on the sixth application. The guy who really wanted to work for a particular NGO got offered a job with them after two years of making relevant contacts and applying for every relevant vacancy that came up. Sometimes serendipity and luck come into play, but staying focused and not getting disheartened is also needed. Everyone has a story about how they landed their first job in the sector. To make your job search process as effective as possible, utilise all the available strategies. This section highlights some of the main techniques but your chances of success will increase as you combine your approaches.

Responding to job adverts

You know the drill: an organisation that has a vacancy to fill – either a new position, if recent funding has been secured, or taking over from someone – advertises the job, and interested candidates apply. You might have to send in your CV and cover-ing letter or fill in a specific form. If the position has been widely advertised on a job vacancy board, your application along with 100+ other hopefuls' lands in the recruiting manager's inbox and from here a shortlist of around six to ten candidates will be made for the first round of interviews. Half of these might be invited back for the second round and one person appointed for the final position. In other words, it's a tough world out there, even more so for the entry level jobs requiring less technical experience.

Many organisations disseminate vacancies more through internet job sites; these are a good first port of call for understanding the types of organisations recruiting and what jobs are out there. It is worth checking these on a regular basis and signing up to their job bulletin email updates if they have this facility. Many of these sites (listed below) also provide an important source for development and humanitarian information, news and good practice keeping professionals connected.

But not all jobs will be advertised on job sites. It incurs a fee and attracts a lot of attention (trawling through 200 CVs can be a lengthy process), so many times organisations may only advertise on their own website or through their own networks. If you have a good organisational shortlist (see Chapter 4) check their job pages regularly.

Often only the high profile jobs are advertised in the national newspapers due to the costs this incurs, but this also varies country by country. Some of the main internet jobsites are listed here. Many of these also advertise volunteering oppor-tunities and internships. Unless otherwise indicated, content is free for jobseekers.

- ★DEVEX is a valuable global resource with an extensive list of jobs (searchable by sector, geographic location and level). It also features news and organisation profiles within the sector. It is free to sign up but some content requires subscription. www.devex.org
- ★CharityJob is a UK-based job site, which has a large number of jobs in the UK not-for-profit sector, international development jobs and volunteer positions. You can sign up for a weekly job bulletin. Content is free for job seekers. www.charityjob.org
- ★IDEALIST lists opportunities working with NGOs based in the US and internationally. You can search by sector and country. www.idealist.org
- ★RELIEFWEB is an important site for the humanitarian sector, with job vacancies, information and analysis. They also advertise relevant training. http://reliefweb.int/
- ★DEVNETJOBS specifically focuses on international development jobs and consulting opportunities. Some content is free, other requires membership. www.devnetjobs.org/InsideNGOs strengthens the 'inside' staff of international NGOs. They have a job board with finance, grants and contracts jobs, also HR, programming, IT and consultancies. www.insidengo.org
- ELDIS from the Institute of Development Studies (IDS) in Brighton, UK, shares information on jobs as well as policy, practice and research. www.eldis.org

- Pambazuka News focuses on social justice issues and is designed specifically for those working in Africa. They produce a free electronic weekly newsletter which also features jobs. www.pambazuka.org
- BOND is a membership body for UK-based organisations working in international development. They list jobs as well as volunteer and internship opportunities and also training courses in their London branch. www.bond.org.uk
- UNJobs shows all the internal and external vacancies available in different UN agencies. www.unjobs.org
- Jobs4Development. www.jobs4development.com
- One World Group also advertises jobs and voluntary and internship positions at http://oneworldgroup.org/jobs
- AlertNet is part of the Thomson Reuters foundation and is a humanitarian news site that also lists jobs in the sector. www.trust.org/alertnet/
- Haces Falta is a Spanish site by Canal Voluntario that advertises jobs and volunteer positions in Spain and internationally. www.hacesfalta.org

(*denotes that you are able you to set up an online profile)

Niche jobs sites, where available, are listed in Part 4 under each speciality.

When searching and applying for jobs the key is to identify the job vacancies that closely fit your skills, qualifications and experience so you can really sell yourself as the ideal candidate for the job. Completing a job application takes several hours, as your CV and covering letter should be adapted for the specific role. Expect online application (where the organisation has a standard format) to require upwards of four hours to complete. Maximise your chances of success by being selective about the jobs you apply for and getting your initial approach (CV and covering letter) right. See the section on CVs below.

Networking

A lot of people think, 'Well, I've put an application in, and I'll leave that to the recruitment gods.' But there's actually quite an important component of active networking so that people are aware of what you do and what your reputation is, and so that we are actually looking for your application rather than leaving it to chance.

UN recruitment chief

Networking is a charged word. Being a good networker doesn't have to mean handing your business card out to everyone at the party, getting theirs, and following it up with an email the very next day to arrange a coffee. In fact, if you overdo it, your reputation could be at stake and it might have quite the opposite of the intended effect. Instead, building up your network is something that should happen naturally, over time, as you work or research in different areas. A trait common to natural networkers is having an interest in people, who they are and what they do, whether they are the CEO or the cleaner.

Networking is so important in the development and humanitarian sectors because a large number of positions are not formally advertised. For an overworked manager, a candidate stands no greater advantage than being physically present, with a known ability to perform the role, or having a personal recommendation from someone close, especially when recruitment is global. In some larger organisations, where HR procedures are so bureaucratic, short-term positions are rarely advertised and often recruited without the human resources department's help. It is less about nepotism and more about finding the right person for the job in the shortest amount of time.

This puts those who have worked in the field for a length of time, with a wide network of contacts, in a better position. But what does it mean to the newcomer with very few relevant contacts? There are some strategies to build up your networks:

- Look close to home. Are there any family friends who work for an organisation of interest to you who might be able to give you some advice or let you know about any upcoming vacancies? A friend of a friend in a high place? Don't feel shy to draw on these connections.
- Arrange informational interviews with your shortlist of favourite employers. Many HR departments are used to such requests and if they have the time will be happy to give out advice (see Chapter 4).
- Attend career fairs. Research organisations before you go and prepare some relevant topics.
- Take part in forums or discussions groups, either physical or online. LinkedIn for example has a number of professional groups such as *International Development*.
- Do some strategic volunteering or internships with organisations that are within your area of interest and/or that are expanding, which means they will be hiring (see Chapter 6, Volunteering and internships).
- If you are already working in an organisation or researching a topic, find out who is working on similar issues, or which organisations have similar remits and connect with them. You have a legitimate reason to connect and discuss. Creating a round table or conference on a particular issue of interest involving experts in the field provides one of the best venues for networking.
- When you are looking for a job, email your close contacts, letting them know that you are available on the job market and asking them to let you know if they hear of something that has come up.

But some words of caution. Be aware of the cross-cultural elements to networking. For example, in the private sector gift-giving is often accepted as part of business,

DID YOU KNOW? TOP TIPS ON NETWORKING

- *Don't expect immediate results.* Networks are built up over time and there are unlikely to be immediate job payoffs, especially if you have few skills to market at this stage.
- *Be interested in people.* Make an effort to speak to people, especially at relevant conferences and events, meetings or forums. It's not just about jobs but getting to know people and their stories, because life is interesting.
- *Remember people.* Do get their business card if possible, or remember their name and organisation. It is OK to follow up with a message to say it was nice to meet them. This is also a good way to have their details on record in future.
- *Connect people.* View networking not as what people can do for you, but what you can do for them. If you know two people working in the same field who should really get to know one another, connect them; if you see a job relevant to someone else, send it to them. This will help them remember you. What goes around comes around.

but it is not accepted in the multilateral sector or NGO sectors and might even offend or be seen as a bribe. Reputation is an important aspect of networking so it is important to manage yours. Get the balance right and network in a non–invasive and sensitive manner.

My advice to anyone on the job search is do not give up, make as many contacts as possible en route and take all opportunities available. You never know where each will lead. I had a Master's in Social Work and seven+ years' experience working in the US when I decided I wanted to switch tracks and work in Africa. I volunteer taught in Namibia and then became a Country Director for a small NGO in Rwanda. With over four years of experience working in the field, I thought it would be quite easy getting an international development job back in Boston. Well, I was wrong. I found the transition back to development work in the US very hard because there was so much competition.

It took me over a year to get my break and it was all down to serendipity and connections. I volunteered for two days a week with an organisation I really wanted to work for, in the hope it might get me a foot in the door, and applied for every single job that came up with them. A colleague there told me about an international development careers fair at a local university that hadn't been advertised but was open to everyone. I went along and it was the best thing I did. Rather than being at a large networking fair, which are fairly common in Boston, this one was small and intimate.

> That is where I found out about the organisation I am currently working for. The girl managing the stand and I hit it off. We had several things in common and eventually met up for a drink. A few months later, she told me about a job vacancy and encouraged me to apply. And I got it! The 20+ applications I sent off in my year in Boston didn't get me anywhere. It was all down to connection and being headstrong. Persistence pays off.
>
> *Jessica (USA)*

Increase your visibility

> When we have a vacancy I look through sites such as LinkedIn, Devex and Assortis profiles in order to identify candidates that have relevant experience.
>
> *HR manager*

Another way to widen your reach is to increase your online visibility and people's access to your CV and experience. Several of the recruitment websites listed above (those with an asterisk) give you the opportunity to create your personal profile where you can detail your career history, upload your CV and highlight your career goals to increase your visibility. How likely is this strategy to land you a job?

You may be interested to know that several HR professionals and recruiters within the field do search for personal profiles on LinkedIn, Devex, CharityJob and so on to try to find suitable candidates, especially for hard-to-fill vacancies that require slightly obscure expertise. This strategy is likely to be more relevant to mid- and senior-level professionals with specific experience. However, this should not discourage others from including and regularly updating their profile. LinkedIn is especially important because people will often turn to it after a first encounter, to find out more about you and your experiences. It is also becoming an important networking tool, allowing you to connect online, view each other's connections, and ask to be introduced. There are also several LinkedIn groups and forums, with several related to development and humanitarian assistance, where issues can be debated and jobs are posted.

Before uploading your CV online, or reviewing your LinkedIn profile, make sure you use key wording to make it easier for people to search for you. Read more in the CV section of this chapter.

DID YOU KNOW? INSIDE THE HR DEPARTMENT: USING LINKEDIN

As an International Development head-hunter, I use social media such as LinkedIn to find and check potential candidates. I am not alone: over 93 per cent of recruiters now use LinkedIn, so you cannot afford not to have a presence.

LinkedIn and Twitter can help you access, connect and interact with powerful people in your job hunt, as well as find jobs. My advice when using social media for networking is first to get your profile right:

- *Separate out your social media accounts* – use LinkedIn and Twitter for professional purposes and job-hunting, Facebook for personal stuff (you can let friends on Facebook know you are job hunting: they may have some contacts and ideas). Employers do check Facebook accounts and do Google searches so make sure they cannot see anything that might count against you – lock down the privacy settings.
- *Ensure that your LinkedIn profile is well written, complete, and packed with appropriate key words* so that you come up in searches. LinkedIn is your online CV and brand. Use a professional photo in formal work clothes (not a holiday snap). Think carefully about the tagline under your name – it must be clear and accurate. Check your spelling and grammar, and ask friends for feedback on how to improve your profile. Update your profile regularly and share news and links so that your name appears on the Update feed. Your LinkedIn profile should be consistent with your CV, though you can add in links and extra information (such as recommendations) that you may not have space for, or may not be appropriate on a CV. A compelling Summary will make you stand out from the crowd.
- *Be open and honest about your career history,* and if you are currently unemployed. If employers need a vacancy filled urgently it is an advantage to know that you are available.
- *Get recommendations on your LinkedIn profile* from former colleagues, managers and partner organisations, particularly donors. Ensure you have a full list of Skills and Expertise so that people can endorse you.

Once you have your LinkedIn profile in order, you can start using it to network. First build up your contacts, and connect with a few new people every day. Aim for a network of at least 300 people, which starts to create a critical mass. The more contacts you have, the more likely it is that LinkedIn will start being useful for you. I have around 5,000 connections. Research INGOs, organisations and individuals you want to target. If you do not have any existing connection to someone you would like to be linked to, you can google their email address, join a group they belong to or request an introduction from a second degree contact. Always write a personal note explaining why you are interested to connect rather than using a canned message.

Do connect with former work colleagues, and let them know you are looking for new opportunities. Join any relevant professional LinkedIn groups, and Twitter chats. In addition to helping you keep abreast of the latest developments and job openings this also has the added advantage of allowing you to connect to other relevant group members in the sector. Also, try to connect with specialist INGO recruiters and HR staff, and follow their Twitter feeds. At Oxford HR we regularly tweet about job vacancies and the sort of candidates we are looking for.

Karen Twining, Oxford HR uk.linkedin.com/in/karentwining

Recruitment agencies, emergency rosters and executive search

There are a number of companies and organisations whose business it is to identify and recruit qualified candidates for international development and humanitarian jobs. These are recruitment agencies, emergency rosters and executive search companies.

Recruitment agencies work with NGOs or government agencies to fill temporary admin jobs with the right candidates as well as put forward relevant candidates for mid- to senior-level professional full-time roles. They usually work from their large databases of candidates who have registered their CV with them. Most work in the national not-for-profit sector, but will be relatively junior in the INGO sector. Some examples from the UK are Charity People www.charitypeople.co.uk; Prospect-us www.prospect-us.co.uk; and The Principle Partnership www.tpp.co.uk/. A temping admin role in the third sector or maternity cover can be a good way to get some paid experience, even if it's not directly in line with your field of interest.

When an emergency strikes, specialists need to be deployed at very short notice, often with as little as 48 hour's preparation. There is no time to start advertising and recruiting for candidates at such short notice and this is where *emergency rosters* come in. Several organisations maintain rosters of people that can be called upon at short notice, providing the UN and other emergency organisations with the personnel they require. If you have a specialisation in one of the functional areas needed in emergencies such as protection, education, nutrition, health, water, sanitation and hygiene (WASH), emergencies, cluster coordination, camp management, supply and logistics, or human resources, in addition to previous experience of working in a development or emergency context you could apply for roster membership. Being accepted on to a roster involves an extensive vetting process, often with interviews about past experience and reference checking. Being placed on an emergency roster does not guarantee that you will be called up for an assignment, but it does mean that when an emergency takes place, you will be among the first to be considered.

> Roster life has its plusses and minuses. Contracts tend to be short, and can't be renewed beyond 12 months. On the plus side, it gives the opportunity to work with a variety of UN agencies, which would be much, much more difficult to do independently. Another plus is the vastly reduced bureaucracy. UN recruitment processes seem to move at glacial speed, so the quick turnaround in the standby roster system lets me bypass all of that.
>
> *Robyn (Australia)*

Some of the organisations that maintain rosters and supply professionals during emergencies include:

- RedR – there are various regional offices; visit the relevant websites to understand the process of becoming a member. www.redr.org/ They also have comprehensive training courses for emergency relief workers;

- UNICEF e-Global Web Roster (e-GWR) – the candidates that are successfully selected for the e-GWR will receive a notification that their profile was successfully registered on the e-GWR;
- CARE International Roster for emergency deployment (RED) www.care-international-roster.org/ – for staff and external consultants;
- Danish Refugee Council – applications to become a member are made within a specific time frame; check out their website for details www.drc.dk/home/;
- Norwegian Refugee Council (NORCAP) maintains an important emergency roster;
- International Rescue Committee (IRC) emergency roster draws on former IRC employees worldwide as well as other qualified external candidates;
- European Union – the European Commission's humanitarian aid and civil protection department, or ECHO, taps its community of experts to provide technical assistance in their respective specialist areas. Currently, nearly 40, all of whom must be EU citizens, have made this list;
- UNOPS which has a roster and their website reflects the jobs they most frequently hire for: project management, procurement, finance, human resources and more.

New relief workers with some experience and relevant qualifications but not enough to qualify for roster work can take advantage of some of the training and support offered by the above organisations. RedR for example offers a mentoring and advice service to RedR Affiliates, in preparation for a career in the humanitarian sector as well as training. The Danish Refugee Council also offers a mentor programme deployment (for new aid workers aged between 25 and 35) www.drc.dk/relief-work.

Executive search ('head-hunting') is less common in the development sector than in other professions, partly because it is a specialist service that is relatively expensive. NGOs cannot always justify paying the 'finding-fee' of these consultancies. A 'head-hunter' will actively look for candidates by asking their contacts for suggestions about who might be interested in the job they are working to fill. They then interview and take up references on 'long-listed' candidates and put forward a handful of pre-selected people for final interview. If you have a number of years' experience in the sector, and are looking for a senior or highly specialised job, it might be worth checking out some of the following agencies and registering your details with them:

- *Oxford HR* works exclusively in the international development and social enterprise sectors around the world. They recruit for board and senior management levels and find talented people to meet their clients' needs. They specialise in finding people for 'hard-to-fill positions'. Vacancies are advertised on their website – apply with CV and completed application form. Clients include CARE, Fairtrade Labelling Organisation, Save the Children, Amnesty, etc. www.oxfordhr.co.uk/

- *Skills for Causes* match people with humanitarian, development or not-for-profit organisations, charities and foundations or companies with a charitable mission anywhere in the world. www.skillsforcauses.com
- *Mission Talent* are based in South Africa, Germany and Kenya. They work to fill leadership positions within international NGOs, large national non-profits, foundations, government organisations, and the UN. www.missiontalent.com
- *Global Recruitment Specialists* is a boutique recruitment firm. With a client base of major development organisations in 2011 they filled 38 positions. They accept unsolicited resumés/CVs for review and consideration. www.global recruitment.net

Other large recruitment conglomerates who work with the not-for-profit sector to help them with their executive search include: Perrett Laver, Odgers, Heidrick & Struggles and Russell Reynolds.

Getting the approach right: CV, covering letter and interviews

Hiring managers only have a couple of seconds to scan your CV and put you on one pile or another: *has potential* or into the shredder. It's worth investing time and effort to get your application right if you want to make it to the long-list stage. This section gives you some tips to get your CV, covering letter and interview preparation right but assumes that you are not new to writing a CV or covering letter. If you are, there are several online and print resources out there to provide more in-depth guidance.

Curriculum vitae (CV)

Your CV creates that all-important first impression. It's important to get it right. What you are aiming for is clarity, cohesion and customisation. If you feel you are right for the job but you don't get shortlisted, perhaps your CV isn't shouting out as clearly as others.

It might take you several days to write or update your CV for the development or humanitarian sectors, but it is time well spent. It is advisable to have a standard CV to hand, to give out prospectively, but when you a spot a relevant job that ties in with your skills and experience that you want to apply for, customise, customise, customise.

When applying for a specific job, customising your standard CV to the criteria of that specific role is one of most important things you can do. The purpose of this is to make the recruitment manager's job as easy for them as possible by highlighting your experience that is directly relevant to the job and mirroring the language that they use. Your suitability for the role will be immediately evident to them – rather than them having to spend several minutes thinking laterally, to find the connection between you and the job.

CV BASICS: FORMAT

1   Presentation counts: make it clean, concise and ideally two pages – although if you have very limited experience, don't drag it out, fit it all on one page as adding padding or pages won't make you look more experienced. If you have a wealth of experience it may be acceptable if it spills over to three or four pages to illustrate this – but keep it to the point. Better still create two versions of your CV, a short one and a long version which has all the details. Let recruiters know that you have a detailed long version, but send your concise shorter CV for initial contacts.

2   At the top of your CV you *can* include a short summary of your key competencies (what you do well) and a synopsis of your past experiences and key achievement (tailored of course to the specific job you are applying for). *Don't* talk about your future plans (career objectives) or motivation in this space unless you are a career-changer, but even then it might be better to do this in the covering letter.

3   The preferred CV format is *reverse chronological*, where you list your professional experience (company, job title and achievements in that role) in reverse order. Then comes education and training and any publications or consultancy work.

4   When giving details of your role in each job use action words such as *coordinated, organised, led, managed, initiated*. When working on projects make sure you include key details such as the size of the budget, the donor, the country and the partners. But instead of just listing your responsibilities, show them what you can do. Summarise your accomplishments with numbers and facts. It's less about what you did, more about what you achieved. You can use bullet points, but keep each one short – a couple of sentences. If it's too long you will lose your reader.

5   Language proficiencies, date of birth, marital status, membership of professional associations, selected publications and countries where you have worked might also be listed at the end. Unless asked for there is no need to include references in your CV.

6   Individual countries and organisations may have specific requirements so be aware of these. In Germany, for example, applicants should include photos in their applications, but in the UK this is looked upon negatively. In some African countries it is common to include your mother's and father's names in your CV, the longer the CV the better and certificates should also be included. Several organisations don't accept CVs but ask you to complete a set application form. The UN, for example, will only accept the personal history P11 form (not be confused with the UK tax return form!) and Europaid uses the Europass form.

7   It goes without saying that you should never lie or mislead in your CV. You can pull out and emphasise some experience more than another, to customise your CV to a role, and draw on transferable skills, but don't ever invent experience or qualifications you don't have.

DID YOU KNOW?

When recruiters use CV databases to find potential candidates they use key word searches. So if you are uploading your CV on to one of the websites listed above (those marked with an asterisk) make sure you include key words in your CV. What are the right key words to use? Think about all the key things that define you. If you were an HR manager, looking to recruit a person just like you, what key words would you use? Make a list and integrate these into a generic CV. *If something can be said in two ways, such as M&E and monitoring and evaluation, make sure that both feature somewhere in your CV.*

In large organisations, HR staff will be given a set of criteria for a job and then a tick box for each candidate on how closely they meet those criteria. They will go through your CV ticking each box. So you want the experience that they are looking for – and that you have – to immediately jump out at them. Expand upon the experience that is directly related to the job, still mentioning other less relevant roles, but without going into these in so much depth. Your CV should basically be answering the question 'why am I the best person for the job?'.

This is what can take time as it involves 1) dissecting the job advert in detail and pulling out all the key words and 2) customising your CV so that it focuses on those particular areas of interest to the recruitment manager.

Different sectors may use different terminology for the same activity, so mirroring the language of the job advert is especially important for sector changers. For example, what some sectors might call training is more often called *capacity building* in the development sector. An overworked HR manager may not automatically make this connection, so it's important to use the same terminology that they do.

Now check, check, check! Spelling errors will land you in the rubbish bin.

Covering letter

Your letter of motivation, letter of intention or covering letter is required for almost all jobs applications. It's a valuable tool for you to again convince the recruitment manager that you are the right person for the role. But don't fall into the trap of repeating your CV in your covering letter. Use it as a unique opportunity to go into further depth or detail about the role.

Your covering letter should answer the following questions:

1   Why am I so interested in this role? What attracts me to the organisation?
2   What particular skills and experience would I be able to draw upon that are particularly related to this role? If you don't have the right experience but do have transferable skills, then the covering letter is the place to demonstrate that.

CAREER CLINIC!

I have worked in the development sector for a number of years, mostly in shorter-term paid roles and voluntary positions and have a Master's in International Development. The problem is my roles are slightly disjointed, one focused on M&E, another on research; I have done some project management and another role was around communications and advocacy. I have a breadth of experience but it is all very limited, so I don't get shortlisted for jobs in any of these areas. What can I do?

You have a range of very broad and relevant experience in development, an excellent all-rounder. Many roles require project management, research and M&E skills. Communications and advocacy are also really valuable skills. You would be an asset to any project assistant or project coordinator role. If you want to specialise in a particular area (M&E for example) you might be better off going for a generic role and then, once you are in, directing your work more towards your particular area of interest (see Chapter 9, Advancing in your career). In your CV really highlight your activities from past jobs relevant to each role and you will find you have a lot to draw upon.

3   If you are trying to make the non-profit leap, or if this role diverges slightly from your previous work, or you are too junior/senior, what are your justifications for applying? For example if you are junior, but a fast learner, and have previously demonstrated this, hiring you could mean that you grow into the role and remain happy in it for longer. If you are senior, but there are aspects of the job you have never done before, there is still a lot of scope for learning.
4   Skills are important but soul also matters. What drives you to do this type of work?

Again, use key words and mirror the language of the job advert as much as possible so that these terms immediately jump out.

Interviews

You have overcome the first major hurdle, now you have a chance to convince them face to face that you are the right candidate. Preparing well for an interview is the key to success. The interviewer is looking for three things during an interview: competencies, character and chemistry. You can prepare the competency side of the interview, and this will give you confidence, make you enthusiastic about the role and allow your personality to come out, and enable you to build up a rapport with the interviewer.

Competency interviews assess how closely your skills match the job. Read through the roles, responsibilities and person specification. What questions might they ask you to test each competency? Prepare a roll-off-the-tongue answer of what previous experience you have in this area, drawing on a variety of different positions and roles. Have anecdotes about past experience ready to demonstrate all the competencies you are asked for.

Examples of interview questions:

- What is your personal five-year plan and how will this job take you there? *Know where you want to go and show how this job meets your aspirations.*
- Why do you think you would enjoy this role/organisation?
- What would you contribute to this organisation?
- We expect part of the role to involve some advocacy too. Do you have any experience in this?
- This role is working in a post-conflict environment. What particular considerations does one have to take into account when running programmes in this context?
- What in your opinion, and from your experience in the field, do you think are the current three main issues in the humanitarian response debate?[1]
- Tell us about an opportunity you've had to network with representatives of similar organisations. What benefits to your organisation resulted?[2]
- Give me an example of when you were involved with the design and implementation of a project. What steps did you follow in developing a sound project plan and what impact did your planning have on achieving the intended results?
- Describe any fundraising successes you have been part of and your role in their success.
- What are three of your strengths and three of your weaknesses?
- Do you have any questions? *Always have some prepared!*

If you have little professional experience to draw on – perhaps you have just finished a Master's and have a couple of internships or overseas time volunteering behind you – don't be afraid to draw on this. They wouldn't have shortlisted you for interview if they didn't feel you had enough experience. You don't need to have managed a $1 million budget to know the project cycle and have been engaged in all of its components. Get familiar with projects, and jargon.

I found myself in Morocco (studying Arabic) when I was applying for jobs in Yemen. Because of the international nature of the job market, many first-round interviews are done over Skype. This is a different dynamic, as you have to build a rapport virtually and overcome technical problems. It is important to ensure you have tested the technology (sound, internet connection) before starting. Remain calm during technical problems, but have a back-up plan if they persist (fully charged phone, telephone number to call in case Skype doesn't work). Never be afraid to ask the interviewers to repeat the questions. And don't show frustration if you have to keep repeating your response. Smile when you talk. It may sound silly but the person on the other end can instantly feel your confidence.

Liny (Indonesia)

Not getting anywhere?

Rejection is unavoidable, so be prepared for it – especially at the beginning when you are lacking experience, but also throughout your career. A popular guideline for the sector is that you need to apply for ten jobs to get invited for an interview, and go for ten interviews to get offered that job. If you haven't landed your ideal job well before you send off the hundredth application, it is probably time to examine what you are doing wrong and modify your strategy. Rejection usually boils down to:

You . . . are applying for the wrong jobs

First, identify and apply for jobs at the right level. From the job description you feel capable of performing on the job, yet may not be given the chance to prove yourself as other candidates stand out more. If the person specification states a Master's and three years' experience, with no Master's and only six months' volunteer work, how do you begin to compete for this? If they are recruiting for a gender specialist and trainer in Afghanistan, your experience of working on a gender policy in West Africa may not be the most relevant. Applying for the wrong level of jobs is one of the most common mistakes people new to the sector make. You have to be prepared at least to start from the bottom and move up, unless you get a lucky break. Those with relatively little experience but niche skills or specialist knowledge are in a better position when applying for jobs.

> I was so surprised when I got the job! It was a manager role and I previously had only done volunteer work and internships. I was told that I was competing against candidates at a much more senior level than me, but they thought I had potential and I did have quite a bit of relevant experience that I could draw on. They are a forward-thinking employer.
>
> *Anonymous*

NOTE: Many jobs have nationality restrictions where you need to already be in possession of a work permit for that country or be a member of an unrepresented member state to be eligible. Check you comply with these criteria.

ACTION: Dissect the job adverts carefully and be selective about the jobs you apply for. First, make sure you tick all, or most of the boxes of the person specification. You also need to be enthusiastic about the job, so apply for jobs you are really interested in and where you can demonstrate how your current experience is transferable to that role.

You . . . have some gaps in your CV that need to be filled

Remember that no one starts at the top. You will be unlikely to get your ideal job at the first go and you will probably have to make multiple strategic job changes to help you to build the relevant experience. You might choose to apply for jobs that you are more likely to get and use the position to gain relevant experience so you can move on to a next role. Once inside an organisation, there are often chances to move up quickly and be able to prove yourself.

> I couldn't afford to volunteer so after I finished my Master's in International Development I went back to my old job in banking. I kept looking for jobs, but didn't even get shortlisted for the ones I applied for. So I decided that, instead of volunteering, I would go for positions that were lower than my experience. I eventually got one in Brussels. Initially it felt like a big step down, but I was able to move up quickly and adapt the role towards my areas of interest. The strategy paid off. Seven years later I was a Country Manager for a large international NGO.
>
> *Maggie (UK)*

ACTION: Apply for jobs strategically. Identify the skills or experience *gaps* between your current profile and the type of jobs you want to go for. Could you enhance your employability by gaining specific experience? You can also recruit the services of a development career coach or get feedback from HR staff when you have applied for jobs (many will be too busy but occasionally they provide feedback). Once you have identified your shortcomings, plan to build up skills in the relevant areas.

Be careful not to fall into the trap of doing multiple volunteer or internships in similar roles, simply because they are easier to get, but don't necessarily build up additional skills. Equally, if you already have relevant studies, work experience is often seen more favourably than a long list of academic qualifications. It is worth persevering until you get a paid position. Widen your target organisations (see Chapter 3) and build up your contacts.

> It is important not to sell yourself short. After you have gained some experience go for a real job, not for another internship. I had one intern here for 18 months, who then went on to another internship. She had gained enough experience to get a real job but internships are easier to get. Be careful not to join an organisation that takes advantage of interns and cheap labour.
>
> *Sian (UK)*

Finally, you may be applying for the right jobs, with the right approach, but have not invested enough in building up your *contacts* within the sector. Disappointingly, many jobs are written with a person already in mind. You also have internal candidates – a known entity – to compete against, making it even more difficult to break in.

You . . . are applying for the most popular jobs (along with 150 others)

Jobs that are advertised on many different job sites are likely to attract many more applicants than those *only* advertised on an organisation's own website, or through direct contacts. When you apply for a job that has attracted 150 or more applications, it will be more difficult to stand out than where there are only 20. The less visible the vacancy, the fewer people that will apply.

ACTION: Develop a comprehensive organisational shortlist (see Chapter 4) and regularly check their website for vacancies and build up contacts within the organisations who can notify you when an opportunity becomes available. Timing is also very important: watch out for organisations that have new funding as this is likely to mean new openings.

> For those looking for their break in the humanitarian sector, timing is important. After my Network on Humanitarian Assistance (NOHA) Master's I spent over a year working for a donor in Pakistan in response to the 2010 and 2011 floods – my job application was in response to a ReliefWeb advertisement that had a very short deadline, and in many ways the timing was fortunate. This is true of finding work in unfolding emergencies, I think. For example, in early 2013, as NGOs were scaling up in Turkey to respond to the Syria crisis, many organisations had quite a few positions advertised looking for people immediately. Being flexible and ready to deploy definitely helps, and organisations with many vacancies to fill may be willing to take on people with a little less experience than usual.
>
> *Anonymous*

You . . . are applying for the right jobs, but your CV lets you down

When you dissect certain job adverts you may feel 'yes, this job is perfect for me!'. You might have the technical experience, regional- or country-specific experience or have worked for similar organisations. In short, you feel this job was created for you. But, disappointingly, you don't even get shortlisted for interview. In this case your approach may be wrong.

Does your CV have a strong and cohesive narrative, making it clear to the recruiter that you are a strong candidate? Do you have a strong enough presence within your community of practice? People recognising your name is a significant bonus and establishing yourself within your field is important.

ACTION: If you haven't tried rewriting your CV in a while, spend some time really working on it to get it right. It can be tedious to do, but crucial to get it right. Give it to some colleagues to review. Are you using key wording and adapting your CV for each individual job? You might even consider asking friends working in the field to show your CV to their hiring department to get some impartial feedback. This is a subtle way to get your CV in front of more people.

If this approach still isn't working, it might be worth investing in the services of a professional CV writer, who is an expert in the development and humanitarian sectors.

One person I worked with at senior executive level had spent over 18 months applying for jobs with no success. We worked on her CV which involved an initial discussion about her career history and the roles she was applying for. Armed with her new CV she got two interviews in the first two weeks.

CV Writer

ACTION: Develop a networking strategy and build up contacts within the field. Organise organisational interviews, asking organisations how they recruit, attend conferences, contribute to discussions and online forums. Get in contact with people to ask for advice.

You . . . are getting interviews, but not being selected for the jobs

Interviewing well is a skill. Practice makes perfect and each failed job interview should be providing you with a learning experience on how to do it better next time. Sometimes it's down to how you answered the questions during the interview, but more often it's about the rapport between you and your interviewer(s).

ACTION: It's important to have prepared well enough so that you are relaxed during the interview, can answer the questions confidently, and ultimately be yourself so that you can build up a good rapport. Practice with a friend or colleagues in preparation for the actual interview. Also check out the profiles of existing employees and interviewees – on the organisation website or LinkedIn. Being selected for a job isn't just about credentials but also *fitting in* and having a good working rapport. Find what you might have in common with them and draw on it.

The candidate I chose to offer the job to was the most enthusiastic; she was full of energy and enthusiasm about the job. We also had a lot to talk about during the interview, and she had some similar experience to mine. She was quite simply the person I felt I would most like to work with of all the candidates. We got along well, and if you are going to be sharing an office 40 hours a week with someone, that is most important. All of the candidates *could* have done the job.

Anonymous

A final word – the job application process is not for the faint-hearted and there are many others who, like you, are applying for the very same jobs. Persistence will pay off and it is not yet time to give up.

Notes

1   www.thecbha.org/media/website/file/CBHA_Core_Humanitarian_Competencies_Guide_Finalpdf.pdf (accessed 30 January 2015).
2   Ibid.

PLATE 5 A coca-producing family – together with another 14,000 farmers – protest against the USA's Plan Colombia to eradicate illicit crops by indiscriminate spraying of their fields. They asked instead for new schools, roads and new employment projects. Most were eventually evicted by force and war brought about by the infamous Plan Colombia.

Source: Ayoze O'Shanahan

PART 3

Moving up in the sectors

You've overcome the initial hurdle – getting your first, second or even third job in the sector. Now you understand more about your field, what the sector has to offer, and where you might want to go. New questions will now emerge as you progress through your career. This section helps you to think about some of the issues that will emerge as you move up or out of the sector.

- Chapter 9: Advancing in your career will help you to move forward in the sector, establishing a direction for your career and taking the necessary steps to get there.
- Chapter 10: Working as a consultant enables you to explore some of the key concerns of consultants, including securing work, money matters and being a good consultant.
- Chapter 11: Starting your own NGO is an introduction for anyone who has an entrepreneurial mind, who feels passionately about an issue that isn't currently being addressed and is driven to fill the gap themselves.
- Chapter 12: Moving on helps you to think about a transition out of the field.

9

ADVANCING IN YOUR CAREER

It's fairly standard within the industry to change roles or organisations every one to three years, although some humanitarian contracts may be for as little as three months. The dynamic nature of development and humanitarian work means that demand is fluid and influenced by funding streams, donors trends and latest issues and development *fashions*.

This lack of long-term stability creates an opportunity for a varied career, but also the need to constantly evaluate the opportunities and develop relevant skills to stay employable, as you will soon be back in the job market.

Some employers do provide a more structured career progression and training opportunities but frequently this will be largely left up to you, especially as organisations are increasingly recruiting for specific jobs, rather than making a long-term investment in a person. You will be in charge of establishing a direction for your career. The more motivated you are and the clearer your goal, the easier it will be to establish the necessary steps to get there. This chapter addresses these elements.

DID YOU KNOW? CHARTING A CAREER PATH

A career path is as personal as a life plan or a value system. It is unique to each individual – it is not given to you, it is something that you 'create'. There are three basic options:

- Wandering from position to position, letting 'fate' control your path.
- Rigidly defining, step by step, the right' path, missing creative opportunities along the way.
- Charting a career path that will help you seek out and evaluate career choices, and recognise opportunities.

Source: UN Career Development Guide[1]

Establishing a direction

Career paths in the sector are not clear and predictable: it is likely you will undertake several quite distinct roles throughout your life. It's difficult to determine a clear direction at the outset as there are no set routes to follow: your senior peers' careers were built in a very different environment. You will come to several junctions in your career, where core decisions will determine your future path. There is a lot to navigate. This first section suggests how you can do this mindfully, to gain some clarity for your future.

CAREER CLINIC! WHERE CAN YOU SEE YOURSELF IN FIVE YEARS' TIME?

This common interview question enables the selection panel to understand your motivations and how the role fits with your career aspirations. But it can be a daunting question if you don't have a clear answer. As a person who embraces opportunities and values unforeseen turnings in the road (flexibility and adaptability are, after all, traits highly valued within the sector), you may have a difficult time – and perhaps feel no need – to answer this question. You figure that your professional identity will slowly emerge as you gain experience in different roles and organisations.

Having a goal and level self-awareness – understanding what you enjoy and can excel at – will help you not just to secure your next role, but ultimately to become a better development or humanitarian professional. Ask yourself: what would you like your contribution to be to global poverty reduction or disaster response? It is important to invest time in this process and find an answer not just for the interview panel but as a crucial step in your career development. There is no need to highlight the exact job and organisation but instead determine your general direction of travel. Once you have reached this level of clarity you will be able to shape all your experiences in this direction, building a solid foundation that will help you to get where you want to go.

Understand and build on your strengths

> Let yourself be silently drawn by the stronger pull of what you really love.
> (Rumi – thirteenth-century Persian poet, jurist and theologian)

Nobody is good at everything, but most of us are great at a few things. It is worth discovering what these are, and investing in what you enjoy. Do you prefer administration and management roles, or are you driven towards a particular speciality and area of focus? Do you like to know everything there is to know about a field, and giving advice? Or do you prefer to make things happen, engaging a group of people into action? Do you thrive in a fast-paced environment or prefer a more iterative organisation that draws heavily on what is evidence based?

Having career clarity from the outset can be very useful, but don't worry, many don't. Experience allows you to understand what you enjoy doing and what you are good at – and equally important – what you don't like doing. If you don't have a clear direction, aim to gain broad experience and take on new and different responsibilities (outside your job description if necessary), speaking to people in the field and understanding what their job involves. Read and research different areas, ideally related and applicable to your current position, so that you can build and adapt these into your practices to gain practical knowledge.

If you have struggled to get into the sector and then find yourself with a job, or sequence of jobs, that are taking you along a particular course, you may be apprehensive about changing routes again. While many of us naturally gravitate towards roles, or are given responsibilities that we visibly enjoy and do well, it is worth analysing this in more detail. If you feel low in motivation, aren't doing as well as you should be doing, it may be because you aren't in a well-matched role or organisation. Do your research, get some advice, and be brave enough to shift course if necessary.

> It's important to find the environment where you can thrive and where your personality, style, mindset and outlook will make a good match with the culture of the organisation. Someone who thrives and feels 'at home' in the culture of UNDP or UNICEF may or may not find the same match at, say, UNHCR or WFP. Even the institutional profile in terms of MBTI type (Myers Brigg) will vary across institutions, some being more orientated to a slightly more strategic/analytical outlook and others to a decidedly more 'operational' one. Both will be great places to be, as long as the match is right. In your career you will reach major forks in the road, and above all it will be important to know yourself, the type of role and environment you can shine in, and to select roles that can reflect your strengths.
>
> *Former UN HR manager*

Keep an eye on the big picture

In the dynamic and rapidly evolving fields that are development and humanitarian assistance – which are increasingly trying to achieve value for money, innovate and work within the evidence base – staying fresh and up to date with relevant skills, technology trends and recent advances is vital. You need to be able to anticipate and adapt to changing demands within your field. This will involve scanning the opportunities, identifying trends and developing relevant skills.

When planning your career, think about the current trends and where they might lead to in the future. What might the 'development' sector to look like in 20 years' time? There may be less aid, and an increased role of the new development actors (the so called BRICS) as well as the private sector and development philanthropists. As aid reduces, there is likely to be a much greater focus on supporting economic development (needing more business skills), governance and effective states (needing more political/judicial/security sector skills). There might be a greater variety of clients you can work for (government/private sector/philanthropists etc.) and what will your role be within this.

Anonymous

One immediate way is to do this is by looking at who organisations are recruiting. This is indicative of where donors are putting their money now, the new projects being funded, trends within the sector and what is in demand. Keep an eye on job adverts for this purpose. Speak to recruiters to understand which positions within your field are difficult to fill and why. An easy way to keep an eye on the job market is to receive some weekly online job digests (see Chapter 8) that you can easily browse in a spare moment. Look specifically at the person specification. What skills and competencies are in demand? This can help you to identify areas where you need to build skills progressively.

Consider specialising (or generalising)

We often think that an efficient way to advance in one's career is to specialise. There is a feeling of quality and performance that goes with saying he/she is a specialist. Specialist skills are in demand, but so too are multi-skilled, versatile and mobile staff who are able to work across occupational groups. By either personal inclination or circumstance you may resemble a specialist or a generalist.

I started out as a generalist until I realised I didn't know enough specifics to be of real use to others in the field that interested me (agricultural business). So I went back to school and got specific, scientific training. As a result I can give more than just an opinion; I can offer solid information and analysis that can help solve actual problems that people have.

Anonymous

The specialist will undertake more technical roles, while the generalist may be better suited to a coordinator or managerial position. But narrow specialisation may limit your opportunities and perspectives and can also result in sudden obsolescence when demands change. At the other end of the spectrum is a generalist, with a range of skills that can be applied in various roles: the quick and flexible learner, excited by a new environment. The risk here is that you may be seen as a 'jack of all trades and master of none', losing out on job opportunities to someone else with more focused knowledge.

In reality these are two opposite extremes of a continuum and most good development professionals are a shade of grey in between. Individuals with well-developed *generalist* skills, combined with one or two areas of *expertise*, are in demand, as are *specialists* who can apply their knowledge flexibly in a range of situations. In the dynamic world of development, the secret is to have agility to adapt oneself between these two extremes.

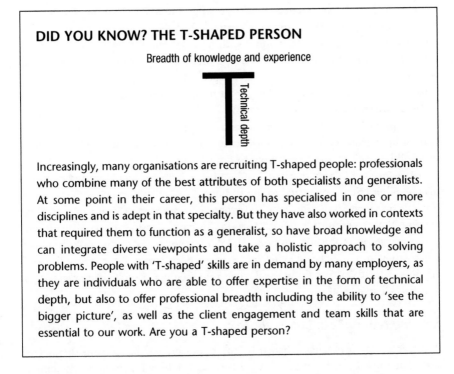

DID YOU KNOW? THE T-SHAPED PERSON

Breadth of knowledge and experience

Increasingly, many organisations are recruiting T-shaped people: professionals who combine many of the best attributes of both specialists and generalists. At some point in their career, this person has specialised in one or more disciplines and is adept in that specialty. But they have also worked in contexts that required them to function as a generalist, so have broad knowledge and can integrate diverse viewpoints and take a holistic approach to solving problems. People with 'T-shaped' skills are in demand by many employers, as they are individuals who are able to offer expertise in the form of technical depth, but also to offer professional breadth including the ability to 'see the bigger picture', as well as the client engagement and team skills that are essential to our work. Are you a T-shaped person?

A word of warning – there can be a tendency, especially in larger organisations, to become an accidental specialist in a specific sub-branch of your field of activity. This can cause challenges because when you are too specialised (in an area that you may discover you don't necessarily enjoy) it isn't then easy to broaden your horizon and try something new or even to carry on doing the things that you really loved in previous roles. If this happens, it's important to recognise it early. You could ask for any training you get to be in these other areas rather than the area you are working on, or try to do a week on a job exchange. Most importantly, don't be afraid to try and find new roles that are matched to broader skills.

> Previously I worked for a small organisation and spent a lot of time with partners to develop campaigning and advocacy that was rooted in our programmatic work. Moving into a much larger organisation I realised that my new role would only cover about 10 per cent of what I previously did, with several different teams – many of whom I never even met – working on the rest. While I devel-

oped very specific expertise, I missed the other broader aspects of my old work. The challenge now is to not be branded as a specialist in this one specific area and make sure that in future job applications my previous skills are still valued.

Matt (UK)

Factor in lifestyle criteria

Development and humanitarian work can offer extremely dynamic and diverse jobs, with frequent travel or living for extended periods in another country. This can be exciting when you are young, with few ties, but if your job is field based, it doesn't always offer a stable existence. Some roles may also take their toll on you, both emotionally and physically, especially where you are submitted to extreme experiences as is often the case in humanitarian emergencies.

Be aware that you may want to move on from a particular role or the sector at some point in the future (see Chapter 12), so it is wise to bear in mind future options in anticipation of this transition. Aim to develop skills in your current role that will be relevant and transferable to future roles.

Salary levels or local benefits such as housing, relocation, schooling and pensions may be other criteria you need to think about. This will determine – to an extent – who you target as an employer and where you direct your career, as there are extreme ranges within the sector. Also, if you have a partner or family you will have to target accompanied positions, and if they too have a career, look for places where they too will be able to get a work permit and have possibilities of employment. If you want a more flexible role, being a consultant may work for you (see Chapter 10).

Set long terms goals, but be flexible

A career in development is shaped by the opportunities that present themselves and your capacity to be exposed to, and engage fully, with those opportunities. Career paths are often defined by chance opportunities to engage in interesting work, taken to the next level. No two development jobs are alike, and you never know what is around the next corner. Despite these unknowns, you do have the capacity to significantly shape this. If you particularly enjoy one aspect of your role, doing additional studies in this area, for example, will enhance your likelihood of being selected for this type of role in future. Try to get the most out of your current role as well. Every job offers the opportunity to immerse yourself in an exciting new field of work; engage with this to the maximum, and learn as much as you can to get the most out of the role.

It is worth taking the time to broadly plot out your career, and some key achievements you would like to reach. Writing down realistic goals bears a surprising correlation with achieving them, and even if your course changes, you can always re-evaluate them.

Career plans must be dynamic and flexible, especially as organisational and personal priorities constantly change. While planning does not guarantee that all your career goals will be accomplished – as there are many factors you cannot control – it does give you a direction. If you don't plan at all, you risk missing out on important opportunities for fulfilling roles by sacrificing your employability.

Getting there

By planning and then carrying out your plans you are controlling the things you can do something about such as your skills, knowledge, performance and qualifications – all of which contribute to your employability and performance on the job. How can you start to implement those plans in your current position?

Direct your job where you want to go

One of the challenges within the sector is that you can't always be selective when it comes to jobs, especially at the start of your career, where the focus is so often on getting a job, any job, within the sector. You may not even be shortlisted for the jobs you really want – and the one you are offered appears to take you off in a different direction. If the job isn't directly aligned with your envisioned career path, fear not. Most people have to make multiple strategic job changes in order to get where they ultimately want to go. Analyse what transferable skills you will gain in it, how it will be viewed by future employers and remember that every job opens up new doors that you probably didn't even know existed.

Once you are in a role, there will be core tasks and responsibilities that you cannot neglect. But there are often opportunities to engage in additional tasks and to develop new skills if you so choose. Think carefully about the aspects of your job that you are good at and enjoy doing. While you may have to invest extra time over evenings or weekends, it is likely to pay off in the long term. Engage your manager as they have an important role in supporting you, especially if you are willing to put in overtime and take on extra responsibilities. Use yearly performance appraisals to identify areas that you would like to develop further. Discuss any opportunities to take on additional responsibilities or tasks. In larger organisations, secondments are also a useful way to develop other skills and gain exposure to organisations, so utilise these if you can. Speak to colleagues who are working in a particular field of interest to you, and ask them whether you can get involved in any particular areas.

> I was working in a law firm and couldn't envision developing a career as a partner. I really wanted to work internationally, but with student loans to pay off I was unable to leave. I then decided to stop seeing it as a nine-to-five job, and instead explored all opportunities in my firm that would prepare me for international jobs. This is how I got involved with the International Criminal Tribunal for Rwanda (ICTR) in Tanzania which my firm has supported through the past ten years on a pro bono basis. Through this I learned about my current full-time international position.
>
> *Jeannetta (USA)*

Get a mentor

> The little child with outstretched arms to an adult is very likely to get carried beyond his/her own height.
>
> (African proverb)

An extremely valuable tool to help anyone on their way to becoming a 'good development/humanitarian worker' is a sympathetic and wise mentor (or network of mentors). Even once you are relatively senior you can derive great value from this, as a mentee or a mentor yourself. Many larger organisations actively encourage and support this and will arrange for staff mentoring, either internal or external. But if your organisation doesn't offer this, you can arrange mentoring yourself: look out for a professional network, action learning set or association that offers this, or approach people individually. Professional career coaches also exist.

Once you have identified a potential mentor in your discipline/profession, check their résumé to ensure their skills or education match where you want to go. Look for someone that you get on well with, as you need to build up a rapport, establish trust and be able to speak openly and honestly with them. A good mentor should also use powerful questioning (asking you questions that are a little bit uncomfortable), enabling you to examine your motivation and direction in more depth. This can help you establish a direction for your career and help you to think about your personal development or simply reflect on your professional practice. A mentor might also work with you to design actions and set goals together.

Many aid and development workers experience stress and isolation. The website whydev.org offers a matching service called DevPeers that connects people at a similar level to professionally support each other. These people can be all over the world and communicate via Skype and email. The DevPeers program helps participants build strong peer relationships, and provides other services to help those working in aid and development be more effective. You can sign up for this service at http://whydev.org. This is what someone who was involved in the DevPeers pilot program had to say about it:

> The fact of having to share with someone who doesn't know you helps you better express/comprehend key aspects about your life, but particularly about your career, as that was our main focus. Listening to myself as well as my peer helped me better understand what I was going through.

FIND OUT MORE!

Mary P. Connor and Julia B. Pokora *Coaching and Mentoring at Work: Developing Effective Practice* (2007, Open University Press) helps both mentors and mentees get the most out a relationship.

Change jobs regularly. . . and don't be afraid of challenges

> To be a manager you don't have to be a technical expert, but you do have to have the ability to learn. Before I start working in a new country, I read and learn everything about the country that I can. I then listen to people and work in teams. I take my time to communicate, and this is so important. Listen to people and work together with them.
>
> *Milton (Honduras)*

The idea of a job for life is not relevant within the sectors. Many people have multiple employers and – probably – several occupations during the course of their careers. Mobility is valued because it means that you are exposed to different organisations, issues, experiences and ways of working that will broaden your knowledge base. It also shows you are adaptable and flexible. It's never good to get too comfortable in a role, chugging on to get your pay cheque. If you find that you can no longer count your achievements at the end of the month and that your motivation is waning, it is probably time to do something about it.

> I have a three-year rule. For me this is the ideal length of time to get stuck into a role, learn the ropes, understand the environment, make an impact, train others and move on. It is a strategy that has served me well. As a versatile manager each new role I take requires me to work in a new thematic area that I don't necessarily know much about. Some might turn them down through lack of confidence. I do everything I can to quickly become an expert in the new field. The body of knowledge and trends isn't usually vast. It is a strategy that has served me well and enabled me to diversify my career into many new and interesting areas.
>
> *Anonymous*

Lateral movement, where you change jobs, but not necessarily level of responsibility, status or pay, is an important career option. A sideways move can offer an opportunity to expand your base skills and knowledge in a particular area, renew motivation or find stimulating new environments. Lateral moves also increase your portfolio of marketable skills and widen your network of personal contacts. Building up global knowledge can be a real asset – transferring what works well in one context and adapting it to another.

> I'm not much of an advocate of career ladders. I think it's important not to view 'rank' or 'seniority' as goals in themselves. I like to focus on job satisfaction. If interaction at grassroots level is what you find fulfilling, then a management role in the country office of a large NGO or UN agency is probably just going to frustrate you. If you really enjoy the sense of facilitating and motivating a team, and creating systems and processes that support that team's effectiveness, you're probably the kind of manager every operation needs.
>
> *Robyn (Australia)*

Adopt a lifelong learning philosophy

> My own personal experience of professional development is that you need to identify your passion – professional stuff you enjoy doing without noticing that it's past midnight – and pursue it.
>
> *Anonymous*

The dynamic landscape of development means that things are constantly on the move as new trends and sectors emerge. Continuous learning and skill development have become a key part of career growth. It is important for people at all stages of their career to identify gaps in their skill sets – and work on these – to be able to stay marketable over time. If you take your continued relevance for granted, you are likely to stagnate.

Learning can be through formal courses (e.g. short courses, diplomas, Master's, PhDs), through personal growth (reading, attending workshops, conferences) or in the job, by engaging in new areas of work, or taking on new roles outside your current expertise. Larger organisations are likely to provide extensive in-house training for staff while smaller ones may have training budgets for their staff. Use these opportunities. But you should also accept that you may have to invest some of your own money and time in your professional development. If you do opt for formal study, the ideal is part-time, online or blended courses. This has never been easier given the mushrooming of e-learning study and short courses.

> If you are in management and want to move up the career ladder, the trend is towards the development–corporate crossover. Higher level leadership positions within INGOs are going to people with an MBA, MPA or leadership training. If this is where you want to be heading, consider completing such a qualification when you are a middle manager, thinking of moving on to a more senior position.
>
> *Hazel Douglas (UK), Executive Search Manager, Oxford HR*

The challenge may be for you to identify areas where you want to gain skills and to envision what lies at the top of your career and how you can get there. There is no framework for the training needs and skills that define a development or humanitarian worker to guide you in your professional development. Speak to your manager and colleagues, and engage with others in similar careers. Professional networks play an important role in the propagation of professional development strategies, innovations and best practices. Some organisations use a core competencies framework when recruiting and to plan staff development and performance.

> Donors and international development professionals are typically *accidental project managers*: professionals with project, programme or portfolio management responsibilities yet without any formal education, background, skills or certifications in the discipline. I, too, was an accidental project manager until

I observed the difference between two projects: one was constantly near collapse and the other was hitting bumps but on track to be successful. The source of this performance difference was the project managers. One was using formal project management standards and the other was not. I decided to self-invest to get myself more skills and credentials in the project management discipline and learned how invaluable they are to project planning and delivery.

Based on an article written by Walter Hekela published in Devex

Keep abreast of developments, share your work, and learn from others

I am passionate about what I do (ICT for development) and spend hours undertaking research on the Internet on ICTs, development and current affairs, which is what helps me to stay ahead in the field. I do have a Master's degree, which is a minimum requirement for many employers. Academic studies can be important but, once you have a degree, move out of the academic fold as I have found that many academics are not very clued up about the development scene. In my area there is no substitute for in-country experience. In fact, you are only as good as the last mission you did! Staying ahead of developments in the field is most important.

Richard (Canada)

The lack of standardised professional development structures means that co-workers are often dependent on each other for learning about innovations and best practices in their field. Professional networks within the field play a crucial role in the formation and dissemination of learning and good practice.

Sharing your work with others and learning about their work is not only of value in itself, but also a great way to continue to build up your networks which will play a vital role in the progress of your career. Some strategies you can employ to do this include:

- actively engaging in both formal and informal professional networks. One starting point is LinkedIn groups, which have a range of discussion groups; you can join *Humanitarian Professional, International Development, Global Public Health* or *Education in Developing Countries*, for example. Research your own field of speciality to find other professional groups and ask others working in a similar field for advice;
- taking part in relevant conferences and forums, not only attending but also offering to make a presentation and speak about your work. This way others can find out about you are doing and it may be the start of something bigger;
- writing about and sharing your work in journals, blogs or magazines. When work is not published it disappears into the grey literature. But many of these reports warrant sharing; there are things to learn and comparison and synthesis

of different consultant contracts in different contexts can be useful. Appendix 1 lists useful reading resources, and many of the journals, blogs or magazines may accept or be interested in featuring your project or work;

- identifying leaders and influencers in your field and following their website or Twitter feeds. Many people use Twitter, for example to disseminate latest news or research in their field. Start doing this yourself too;
- bringing together roundtables of people working on similar issues. Collaborative working often brings about project success. Action learning sets are a good way to discuss and address problems and learn from others working on similar issues.[2]

> As a Master's student and young professional I got the chance to work at the university where I graduated, and I specialised in environment and development. I published a book and some articles based on my experiences in the field. The essential message of my research and development work was that environmental problems are man-made socio-economic problems and can only be solved by people. This was a relatively new idea back then, and the book and articles were widely read. This gave me a feeling of belonging, being part of a professional development community, where my skills added value in a technician-dominated context.
>
> *Ted (Netherlands)*

Be effective in your work

People can be so caught up in their daily work and keeping things ticking over that they don't always have time to look at the bigger picture. What have your major achievements been to date? What has your contribution been to the field? What would you like these to be in future? Remember why you went into the field in the first place and what your ambitions were. Are you just focusing on numbers and outputs, reports to donors and getting funding in? Or can you see the real impact of your work, with the long-term outcomes on the ground?

Purpose and passion – accompanied with the right hard and soft skills – and a motivation to strive to do things better, question the status quo and innovate to bring about lasting change will help you to become the type of leader the development sector needs. Set yourself performance indicators, individually or jointly with your manager, so you have a benchmark to measure whether you are achieving or not.

> I was the Country Director for a large INGO when the 2012 food crisis hit. I needed to act quickly, knowing that my decision and choices were critical to help save people's lives. In these circumstances it is vital to have a very good understanding of the situation, be able to make decisions quickly, motivate the teams and use the institutional knowledge to overcome all the admin hurdles. It was a stressful time, where we were working 18-hour days and we

had to dedicate ourselves fully to the mission. When lives are at risk you don't feel you can rest. My family with three small children was with me but I couldn't find enough time for them. In 2012 I was very proud as I received an MBE Honorary Member of the British Empire from Queen Elizabeth II, an award that I received for my services over this time.

Mbacke (Senegal)

Standing out and being effective in your work, whatever you are doing, will be visible to colleagues, managers, partner organisations and the wider network of stakeholders working in your field. If you want to move forward, strive to be a leader in your field and make change happen. This doesn't always mean that you will always do things well, but try to learn from your mistakes, and share this learning.

We did a housing project in Sri Lanka after the tsunami, which had seemed a successful programme. But when I went back a few years later to do an evaluation, people didn't seem so happy. Why? Sri Lanka, like India, has a caste system, which we had missed entirely during the planning. School teachers were living next to fishermen, which was completely unacceptable to them. It is important to understand what is going on from a cultural aspect, because overlooking this causes many projects to fail.

Milton (Honduras)

And finally, maintain a sense of balance and purpose

If you want a long-term career in the sector, your well-being is important. Long hours, little support and workaholism does not benefit anyone in the long run. Becoming cynical will also affect your motivation and work performance. This can be especially true for people working in a fast-paced humanitarian emergency sector where family dynamics and your health can suffer, but it is relevant to everyone. So how can you take a step out of the highly stressful environment you may find yourself in to gain perspective, and maintain balance and purpose?

CAREER CLINIC!

I have spent several years preparing myself for my 'ideal job'. Now I have it, I don't like it! What can I do?

Many people enter the sector with a great amount of hope and idealism, only to become jaded when they feel they are not able to make a significant contribution, bureaucracy is crippling or programmes are not being implemented as efficiently or logically as you would do it. Others may have been enamoured with the *glamour* of the profession, focusing much more on the organisation rather than thinking about what they would really enjoy doing. It is also not uncommon to be overloaded with work and feel you don't have the required knowledge, information or time to do a good job.

At the start, don't flee unexpected challenges, as every task – even if it wasn't what you expected to be doing – offers a learning opportunity. Don't make snap judgements either as there is usually a reason why things are done the way they are – but this doesn't mean you can't try to change them. Identify the problem areas or fields where you need to gain additional knowledge to do a good job and how to get there. If a bad boss is the issue, turn to colleagues to see how you can work around this. Give the position at least 12–18 months so that you can still put this valuable experience in your CV before considering your next move.

What next?

What is written down is much more likely to be achieved. The worksheet in Figure 9.1 should help you to capture some of the thoughts in this chapter and put your goals down on paper.

Notes

1   *Career Development Guide*, UN Staff Development Service Division for Organizational Development Office of Human Resources Management August 2007 www.un.org/staff development/pdf/CareerGuide070823_low.pdf (accessed 30 January 2015).
2   *Tools for Knowledge and Learning: A Guide for Development and Humanitarian Organisations.* ODI Toolkit, 2006. Available as PDF from www.odi.org.uk/sites/odi.org.uk/files/odi-assets/publications-opinion-files/188.pdf (accessed 30 January 2015).

WORKSHEET: Chapter 9. Setting career goals and reaching them

- Do you have a five year plan? Where would you like to be? What is the vision for your career? (If you have difficulty completing this now, keep going through the worksheet and you can come back to it later.) What would you like your contribution to your sector to be?

_____

_____

_____

then come back and review your five year plan

- What are your main professional experiences, skills, strengths, competencies, knowledge and attributes? List 10.

  - _____
  - _____
  - _____
  - _____
  - _____
  - _____
  - _____
  - _____
  - _____
  - _____

- How do these relate to the T-shaped person? Which are professional breadth and which relate to technical depth?

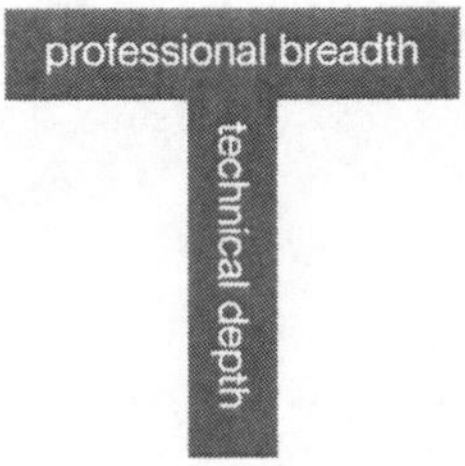

  professional breadth

  technical depth

- What are the trends in the development/humanitarian sectors and which skills and experience are increasingly in demand?

  - _____
  - _____
  - _____

  Think also about your lifestyle criteria. What is important to you?

FIGURE 9.1A Worksheet: setting career goals and reaching them.

Now you have a five year goal, how can you get there?

• Reflecting back on your own goals, current 'T' skills and trends in the development sector, list five (or more) skills, knowledge, competencies and experience which you feel you need to develop.

1. _____   _____

2. _____   _____

3. _____   **Why?**   _____
                                What is the
                                benefit?
4. _____   _____

5. _____   _____

• How will you achieve each of these? Think about ...

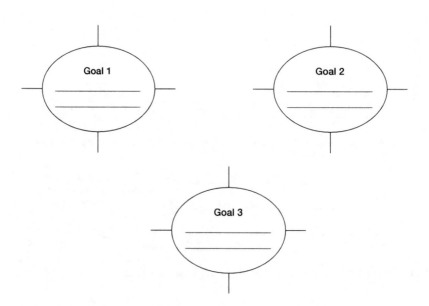

| Getting a mentor | Expanding your current job to new areas | Personal research and learning | A new position closely aligned to future goals |

• Think about how you can use these strategies to reach each of your above goals. What else do you need to do?

Goal 1

Goal 2

Goal 3

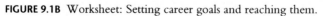

FIGURE 9.1B Worksheet: Setting career goals and reaching them.

10

WORKING AS A CONSULTANT

At the moment I am on a 20-day mission conducting a market analysis for an NGO in Mozambique. Last month I was working on a value chain analysis for fisheries and then palm oil in Ghana. Before that I was the team leader for a UN fact finding mission in Chad. Later this month I will be working on some home-based research assignments and then hope to take a few months off over the summer.

Agribusiness Consultant

When an organisation doesn't have the expertise or capacity within their own staff base, outside experts may also be bought in to support implementation of a project, deliver a specific piece of work or provide an objective outside perspective. Due to the project nature of development work and the short timeframes of some humanitarian assistance roles, there are many consultancy opportunities for those possessing the required knowledge and expertise. Individual assignments may be anything from a few days to those who are deployed on multi-year assignments.

The life as a consultant can be an exciting one, ideal for those who like diversity and new challenges, with projects and clients changing constantly. This is one attraction but also the challenge of the type of work. A consultant can work on a freelance/self-employed basis or work on the staff of a consulting firm, with the accompanying employment benefits that come with the latter. This chapter mostly deals with the first category and covers the transition to consultancy, the work you might engage in, money matters, how to deliver quality work as a consultant and transitioning out of consultancy.

Becoming a consultant

For some, the decision to become a consultant – rather than being a staff member – is a conscious one. They may be searching for a more flexible and diverse lifestyle, being able to pick and choose projects to work on, engaging with a range of organisations and having the luxury of taking chunks of time out. It may also be a natural progression along someone's career, for example when they have developed expertise that is too specialised for an organisation to employ someone in a full-time position, but their skills are in demand on a short-term basis.

For others, consultancy is a sensible temporary employment option: while searching for a full-time role, perhaps after taking some time out for studying, or when a full-time position has come to an end – filling a gap between two jobs. Finding that they enjoy this work and are successful at securing projects, some never go back to *being a full-time employee*.

Some people even manage to combine both, taking time out from a full-time position to take on the occasional consultancy assignment. This is common for people working in research, academia, or for NGOs.

> I did not want to be an accidental consultant: I have seen many organisations reduce their permanent staff base and move to outsourcing certain services, and it is often those who have just been made redundant who are the best qualified for these jobs, so they come back as self-employed consultants. I knew I would like to become a consultant but only after I had gained enough skills, knowledge and confidence to sell my services. I planned my transition to consultancy over a ten-year period, gaining skills, expertise, knowledge and contacts.
>
> *Pamela (Kenya)*

Working as a consultant is no easy task and, before making the leap as a *career* consultant, make sure that you are ready. Not only do you have to deliver quality work, but you also have to *secure* the work in the first place. You are selling your expertise, not just your time, so having significant knowledge and experience in your field is important. The list of tasks that you will have to do is endless, and many of them will be unpaid – strategic planning, writing proposals, developing training materials, answering the phone, marketing or selling your services, filing, writing articles, making tea, preparing accounts and finances – each responsibility is yours alone. You will need to have strong technical skills, (usually) a good graduate education, strong professional networks, experience – preferably in multiple settings – as well as the ability to work independently, and maintain motivation and creativity at times of uncertainty. Building and maintain your reputation is crucial to your success.

> As you are running your own business, every minute counts. I often get invited to lengthy meetings, and you can tell who there is paid a salary, as they are at every meeting and attend the whole time. However, as a consultant you need to pick strategically what meetings will lead you into new work opportunities and choose only those to attend. I will even show up for certain times, i.e. networking or lunch to make the most of my time.
>
> *Debora (Canada)*

The work

The work of a consultant may be anything from supporting an organisation to write their gender strategy, running a training workshop in participatory rural appraisal, conducting a project evaluation, writing a toolkit, managing a piece of research or building a database to managing an entire project. You could be undertaking many of these roles as a staff member, but as a consultant you have the luxury of focusing exclusively on the task in hand and also bring in expertise and an impartial viewpoint. The decision for a role to be a consultancy role or a fixed-term staff position will be an internal one. Some consultants are field based, others do home office work, but most roles will require a combination of both.

On short-term assignments you will have very clear terms of reference and be expected to deliver exactly on these, with little scope for mobility. On longer-term placements you will have a greater opportunity to shape and provide your opinions, influencing what is delivered.

> I am fairly unusual in that I have done very few short-term consultancy positions but mainly long-term project work. Most people in this business have in fact managed very few programmes and have spent most of their career telling others how to do it. I am one of those dangerous types who know a lot about many things but not very much about anything in particular. On the flip side you can be an expert and corner your small niche. There are two agricultural consultants that I admire who are never out of work. One was the world's expert on two types of viruses that infected cassava and the other was the world expert on the lethal yellow disease of coconuts.
>
> *Tim (UK)*

Securing work

Many people get their first consultancy contracts with organisations where they are already known: either as a staff member or a contact. Once they have undertaken a few consultancy assignments and have built up a reputation, they are in a good position to market themselves to a broader audience.

Your ability to secure new work will have a lot to do with how well you market your product: you. Having confidence in your skills and expertise is a must. Contacts play a very important role as does your personal reputation. While some consultancy posts are openly advertised, many are not: it is hard to justify an elaborate recruitment process for a small contract. In fact many consultancy roles never go through the HR department, with project managers hiring directly, as they need to find the right person as quickly as possible, so rely on recommendations. This unfortunately doesn't leave much space for newcomers. Networking and collaboration is crucial for successful consultants. Consequently, many consultants invest a lot of time building up their networks between jobs, on LinkedIn for example, or writing articles or blogs about their work, or contributing to online forums to raise their profile.

Open applications or tenders are usually advertised via the same outlets as job adverts (see Chapter 8). In addition to submitting your CV, clients will often ask you to submit a technical proposal (how you expect to carry out the work), including time frames and a budget (see later section on *Money Matters*). Some consultancy firms will also employ freelance consultants for specific assignments (you can often register online with them). Some will also advertise for consultancy roles, often for a tender that is being submitted, before the funding has been secured.

Once you secure a contract with an organisation, building strong relationships between you and the client and producing high-quality work will ensure that you get called back or get a referral to another organisation. Many consultants find themselves with a few major clients, repeatedly being called back by the same organisations. If you deliver a successful assignment, the firm may not want to lose you and offer you a staff position within the organisation.

> I make a point of going to one international conference in my field at least once a year in a place where the decision makers are. You have to be strategic about it. So I go through the participants list, circle all the names of people I want to meet, and spend my time looking for opportunities to speak to those people. It's more subtle than a sales pitch, but could be getting a coffee together and discussing opportunities. I would say I generate about 80 per cent of my new work through these meetings.
>
> *Debora (Canada)*

Working environment

If you are home based, this will also become your office. This can be great at first, but when your daily routine involves getting out of bed, walking straight into the office and the highlight of the day are your trips to the kitchen you may start to miss office life. With no colleagues to bounce ideas off in the room next door, you are entirely on your own. For people used to working in a team, or having a personal assistant to hand tasks over to, this can be an initial challenge.

> Running my own practice means working alone and from home. Instead of having my colleagues to talk to before making a decision, I find myself staring out through the window talking to the hare in the garden! I walk to the kitchen several times a day to make tea or pinch a piece of cheese. Not leaving the house has been one of my biggest challenges. As a way to get me out, my husband and I have started going to the gym together.
>
> *Pamela (Kenya)*

Some people thrive on this flexibility, and make a roving office in cafés and libraries, while others may rent a desk in a shared office space to separate their work and home life. Some organisations may offer you a desk to work from within their offices, especially if you are field based. But there still may be many times in between jobs, when you are securing new work, that you are based on your own.

In between assignments, your time will be unpaid and you may be anxious about not getting enough work. Not knowing where your next pay cheque is coming from brings about uncertainty and financial instability. You need to stay motivated, believe in yourself, and dedicate the time in between contracts to securing more work and reputation, as well as advancing your own skills and updating your resources.

Once you get an assignment – or two that come in at the same time – you may well be on the job seven days a week with no break if the deadlines are tight. When travelling, working conditions can be difficult, intense and tiring. It is important for a consultant to maintain a sense of purpose and balance, factoring in, and honouring, family and resting time, to ensure you stay fresh and motivated, avoiding illness and burnout.

Money matters

The common perception is that consultants are overpaid, and it's not hard to see why when daily fees for international consultants can be in the region of €350 to €600 a day or more.

It can be difficult knowing what to ask for and how much to negotiate. Speak to other consultants to get a guideline of acceptable daily rates. You should also determine what you would like your gross yearly salary to be (for example €50,000). Then estimate the number of paid days you are likely to be able to secure that year. A very successful consultant may get around 200 paid days of work, but it is not uncommon for a consultant to get only 80 or 100 days' work. If you estimate it at 100 days' work, you need to charge a daily rate of €500 per day. The other 265 days of the year will by no means be a holiday (although you should make sure to take some). If your work involves travel and expenses, check that this will be covered separately and not coming out of your daily stipend. Pension payments, insurance costs, training and other fees will often have to be covered directly by you, so these are other costs that you may also want to factor in.

> When you are starting out as a consultant you may have to take on some unpaid work to gain experience and build up your CV. The challenge is that the organisation may not value your services as they haven't ever paid for them. To avoid this, whenever I do any unpaid work, I take it as a pro bono, where I tell them how much the work should cost or even provide them with an invoice with the value of the service clearly identified on it. The next time the organisation will be more likely to pay for the work.
>
> *Anonymous*

Another challenge is that you may not get a regular pay cheque and you will need to have a reserve so that you can adequately cover your costs for the months in between payments. Contracts can be such that you get a lump sum upon delivery of the final contract, or two instalments, the first after the inception phase. Some organisations can be very slow at paying up while they approve the quality of your final work. It is important to keep clients constantly updated with progress, and any challenges that will not allow you to deliver items on the terms of reference (ToRs) so that there are no surprises at the end, when it comes to payment. You will also need to spend some time doing your accounts, or hire an accountant. Keep a record of all your business-related expenditures and receipts.

A distinction is often made between local and international consultants. Organisations may prefer to hire a local consultant where possible, with a physical presence in the country, saving on costs (daily rates are usually around half what international consultants get paid) and bringing with them existing local cultural understanding and knowledge of the work environment. In countries such as South Africa, Kenya, Ghana, Nigeria, Côte d'Ivoire, in Africa, for example, or India, Indonesia and the Philippines, in Asia, local consultancy capacity is very strong. Here teams are often led by regional or national rather than international consultants. One difference between these emerging cadres of national as opposed to international consultants is their degree of 'global' experience.

Being a *good* consultant

The reputation of consultants and consultancy is not uniformly high. Some *successful* consultants – who are good at securing work – end up taking on too much work, or work that they do not have the right skills for, and are not able to deliver upon the ToRs.

> I have been contracted on numerous occasions to finalise work started by another consultant. On some occasions this is because the work that has been completed is substandard, so a report needs amending and editing and gaps need to be filled. On other occasions a consultant has led a survey or evaluation and the data has been collected, but the consultant has failed to deliver the report. This is a nightmare for the project managers as it delays the whole cycle of work and costs more. Once you become known by an organisation as someone that delivers high-quality work on time you will be called back time and time again. The secret is finding an organisation that has a lot of work to offer you.
>
> *Maria (UK)*

Some organisations place a strong emphasis on academic qualifications when recruiting, while other emphasise a competency framework. Hard skills and technical knowledge are important but a good consultant also brings with them a range of soft skills too. Advice and recommendations need to be appropriately contextualised. The ability to write plainly and concisely is also very important as many clients may not be English speaking and there is a need to stay away from technical development-related jargon. Good consultants understand the art of listening, are able to immerse themselves in the local people's realities as much as is possible and are able to provide a sustained engagement with them (not just 'fly in/fly out') where required.

DID YOU KNOW? SOME QUALITIES OF A GOOD CONSULTANT

- Ability to listen and observe
- Good writing and analytical skills
- Ability to see things from different perspectives
- Ability to employ a variety of research tools and techniques
- Ability to deal with different cultural contexts and relationships
- Adaptation to the specific context of each piece of work
- Flexibility, given that schedules and situations may change, even with the best planning
- Honesty about what is not feasible, and what they cannot deliver
- Persistence in questioning the consultancy brief or ToR in order to ensure a shared understanding of the task
- Timely reporting
- A willingness to check and discuss any required changes
- Empathy with the client organisation

Source: *Effective Consultancies in Development and Humanitarian Programmes*, Oxfam Skills and Practices, John Rowley and Frances Rubin (2006).

Being a *good consultant* also comes down to personality and the relationship between you and your client. Maintaining regular communication flows, and establishing dialogue right from the start is key. As a consultant, it is your responsibility to make sure that you have the rights skills and ability to perform well, and – while it can be difficult – highlight any areas of competence that you may be lacking. All too often the person specification for consultants is over-optimistic. Any gaps that you have may be overcome by hiring in someone else with complementary skills, so that you can work together as a team.

At the outset it is also vital to ensure that everyone has a shared understanding of the ToRs. For example, if you are asked to assess innovative types of work, you need to be clear what is meant by this in order to make sense of the task at hand. Search for clarification where the ToR is vague.

> One of the challenges that we find when recruiting for Chief of Party roles for funders such as USAID is that there is an over-emphasis on technical skills. Some positions are so specific that there might only be four people in the world who would qualify! But the reality is that with these multiyear assignments, the skills needed for the start-up implementation and close-out phase are completely different.
>
> *Chris (USA)*

As a consultant you will often be called upon for your advice or recommendations. Commissioned work may feed into a business case or impact upon decisions further down the line (which as a consultant, you will be unlikely to know about or see). It is important that your recommendations are grounded in evidence-based practice (using and citing the appropriate hierarchy of evidence) and provide a balanced view (not only providing evidence to justify a particular position). You may also be expected to do a cost analysis of the proposed interventions with a comparison with alternative approaches to ensure value for money.

One of the negative associations with consultants is that the knowledge follows them, so when an assignment comes to an end, and the report is submitted, the knowledge is lost with the consultant. You should work to overcome this, agreeing with your client to provide a feedback workshop to ensure cross learning.

Career progression and professional development

> I recently attended the International Program for Development Evaluation Training (known as IPDET) in Ottawa, Canada and there were 170 development professionals from 80 countries. There were participants from donor organisations, NGOs, research institutions, independent consultants (like myself) and from financial institutions (World Bank, Inter-American Development Bank (IDB), etc.). The networking opportunities at the event and number of social activities were just as enriching as the content itself.
>
> *Brenda (USA)*

A challenge of being a consultant is that there isn't always much opportunity for career progression. You will get known for doing a certain type of work or country-specific expertise and keep being offered the same type of work. In order to stop yourself becoming stale in this dynamic environment, and constantly using the same materials and methods, it is important to invest in your professional development to build up your skills set.

As a consultant you will need to take responsibility for your own professional development and pay for this yourself. This is difficult, as field-based project work is usually very demanding and allows very little time for this. Your goal should not only be enhancing your employability and taking on new areas of work, but also working to raise professional standards. Do set time aside – ideally at least ten days a year – in between assignments to attend training, workshops and conferences. You may also want to consider an online training course.

> As a consultant, you are selling your skills. So you first need to ask yourself 'Where can I add value to this organisation, what skills do they need and what are they willing to pay for?' Also once you have identified your focus you need to continue to build up skills in these areas. That may mean additional training courses, certificates, even a Master's.
>
> *Debora (Canada)*

You should also take time at least once a year to review your work and understand which assignment you enjoyed the most, and continue to seek work in these areas. Also ask for feedback from your clients and learn from this. Actively engage with professional networks, both for support and advice, but also to understand new areas of practice and innovations in your field.

> The sphere of one's consultancy work usually changes over the years. Consultants need to be adaptable and responsive to changing demands, willing to undertake their own in-service training and education, and willing to move into fresh fields of professional work as the need and demands arise and their own experience and qualifications develop. My area has always been education, and through my career I have moved through teacher, lecturer, researcher, project manager, project design, monitoring and evaluation.
>
> *Geoff (Australia)*

Transitioning out of consultancy work

The first few years as a consultant can be the hardest, while you are building up contacts, clients and your professional reputation. You may have to deal with a significant amount of insecurity as well as a steep learning curve. You may find that you enjoy the freedom from institutional politics and the ability to focus entirely on one piece of work, in which case, you may decide to carve out further your career in consulting. You may, however, miss being part of an institution, having the support of an organisation and working closely with colleagues. In this case you may decide it is time to transition back into a full-time role.

Some consultancy firms may offer their good consultants a staff position so as to not lose them to the competition. But it is also possible to apply for and work in staff positions for the organisations you have been consulting with. Especially where you have built up contacts and a reputation, having a period in your CV where you have undertaken consultancy work should not count negatively against you.

> As a consultant you are often working to tight deadlines with clear outcomes and producing stand-alone pieces of work. When transitioning back to a full-time job, it requires a change in mindset. This can be difficult at first.
>
> *Maria (UK)*

PLATE 6 Entrepreneurship and social innovation are vital to unlock growth and economic inclusion in developing economies. Entrepreneurs like Henri Nyakarundi in Rwanda play an important role in job creation for youth. With his mobile solar kiosk he is bringing renewable energy to off-grid communities. The stations offer mobile phone charging, solar electricity and, in future, internet hotspots. This low-cost franchise business model offers opportunities for rural entrepreneurs who want to be in business for themselves but not by themselves.

Source: Tom Gilks

11
STARTING YOUR OWN NGO

Entrepreneurial minds interested in working in the international development sector will already have toyed with the idea of setting up their own non-governmental organisation (NGO) – or social enterprise. If you feel passionately about an issue that isn't currently being addressed or see a niche with potential you might be driven to fill the gap yourself.

Many NGOs that dominate the sector today started as an idea that lingered in someone's mind long enough for them to do something about it. Bringing other people together on their vision, they made their first tentative efforts to improve the status quo. Could you be the holder of the next big thing in industry? But like businesses, only a very small number of those that are started make it big. Growing an organisation is a massive undertaking that requires commitment and personal investment. Don't raise hopes and create empty promises that you then can't deliver upon.

This short chapter can only introduce you to the topic, highlight some of the main things to think about when contemplating the set-up of an organisation, share other people's experiences and point you towards other resources that will help you further.

Six things to think about before/when starting up an organisation

There are many similarities between establishing a new organisation and implementing a project. The project cycle is a widely used structure in many development organisations including NGOs, World Bank, UN, etc. It is just one approach that you might find useful in the planning stages of an organisation. Figure 11.1 shows an adapted project cycle, and this chapter guides you through each of these potential stages. Even if your organisation develops more organically, you are likely to be thinking – conscious or subconsciously – about each of the elements.

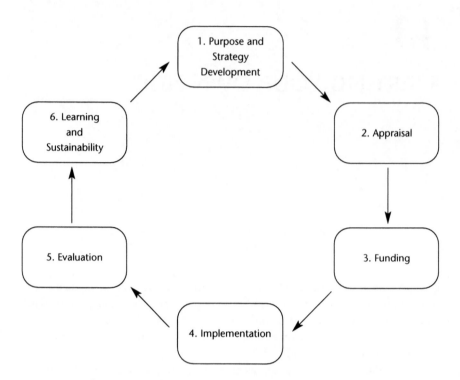

FIGURE 11.1 The project cycle adapted to an NGO set-up.

Purpose and strategy development

Your organisation requires a purpose, direction. You will need to develop a powerful vision and mission statement, conveying what you are trying to achieve, how you will do this and what will help to set your direction. The vision statement spells out the high-level goals while the mission statement is more concise – What do we do? How do we do it? For whom do we do it? This should enable you to motivate and engage others in your work and also help to develop contacts with other organisations that can be built upon once the NGO has been established.

But how do you establish this goal and your organisation's role within it? You should not impose ideas but instead implement solutions. In order to be able to draft the vision and mission statements you need to work closely with your beneficiaries, and see them not as recipients of your help but as partners in the process. Participatory approaches are very important.

Feeling compassion and a desire to do something is not enough: problems such as illiteracy or child labour are complex and deep-rooted, and the effort to address them brings with it a whole other set of challenges. In order to do this effectively you need to have (or be prepared to gain) an in-depth understanding of the context, what your beneficiaries' needs are, who is already working in these areas and what gaps you can fill. Failure to do so is likely to result in your good intention having, at best, no impact, or at worst, a serious negative consequence.

> I was a trustee of a small organisation. It was started by a friend of mine who wanted to 'do something to help' and, after reading an article on AIDS orphans, decided to set up an orphanage. He personally donated a lot of money but also raised money from friends. The orphanage was built and a director recruited with yearly transfers to cover salaries and the children's maintenance. As a trustee and someone with international experience I decided to go and pay it a visit. It had been running for over ten years. In the reports all looked good, very positive in fact. But delving deeper I found that many of the children actually had families, but claimed to be orphans as at the centre they received new clothes, good food and their education was paid for. This had initially created resentment among the local community and some of the real beneficiaries were actually forced to leave. This is the problem of starting a project from afar, when you don't understand the context.
>
> *Anonymous*

Most founders of successful NGOs or social enterprises find the passion and energy to start up an organisation from trying to solve a problem that they understand personally. The most successful NGOs are started by people who know the challenges of the environment and have exposure to the intended beneficiaries, engaging them in the planning stages. Many founders did not start out with the intention of setting up an NGO, but stumble across an innovative idea or need that they feel passionate about as they go about other activities.

> After spending some time in Hanoi's famed Old Quarter, I began understanding some of the social problems that lay beneath it. I probably picked up on these quickly because I know first-hand what it feels like to draw the short straw, growing up poor in Australia, and having my future options limited. I used to sit drinking coffee and talking to the shoe-shine boys. There are tens of thousands of street children in Vietnam. Most find their way to Hanoi or Saigon from poverty-stricken rural areas and are either orphaned, abandoned or sent to earn money for their families back home. I became their impromptu English teacher, and the boys proved to be quick and eager students. So eager that our class roll soon doubled, then doubled again. Then my landlord complained about the inordinate number of 'vagrants' in her house. So it was time I either got serious and started up an NGO, or I moved away.
>
> Ten years later our annual budget is about $1,000,000. Half of this is derived from personal donations; other NGOs and some foreign governments donate to us too. We have a drop-in centre in Hanoi for 250 kids, two residences for high-risk children and teens, a back-to-school programme in the province of Bac Nanh for 800 pupils in danger of dropping out, a legal advocacy service, and programmes in Hue and Dien Bien for children we rescue from trafficking

and slavery. But I worry we are still not doing enough. If you ask me how to start an NGO, I'd say 'If you're prepared to roll up your sleeves and be in it for the long haul, then just start helping people and work out the plans later.'

*Michael Brosowski (Australia), Blue Dragon
Children's Foundation, Vietnam*

One of the biggest challenges after the Indian Ocean tsunami in 2004 and in Haiti in 2010 were the large number of well-meaning NGOs who descended on the area with pockets full of cash trying to help, but in most cases contributing to the chaos on the ground. Within the humanitarian sector best practice standards have been established to regulate and encourage professionalism with organisations working in the field, and include the Humanitarian Accountability Partnerships (HAP International); the Voluntary Code of Conduct for the Red Cross and Red Crescent Movement and NGOs in Disaster Relief; the Sphere Project and the guidelines set out in the *Sphere Handbook*, which organisations should adhere to. Unless you have a strong understanding of the humanitarian field, you are strongly discouraged from setting up an organisation in these contexts.

DIY [Do it yourself] aid is a huge trend created by the massive media engagement in a disaster; everyone watches the pictures on television and the global goodwill pours out – incoherent, passionate and convinced it can make a difference.

*Madeleine Bunting, 'How can we curb the proliferation of NGOs in
a crisis?' Extract from www.guardian.co.uk/global-
development/poverty-matters/2011/may/
11/curbing-proliferation-of-ngos-in-crisis*

Appraisal

In an ideal world, NGOs would not exist, as the state systems would provide all the services and support that their citizens require. This rarely happens, so NGOs are there to fill this gap. If you are using public funding you have the responsibility of ensuring that this is used effectively and efficiently. Once you have a draft vision and mission, understand what other organisations are doing in this area, and what your organisation will add. Don't duplicate, but instead join forces to improve efficiency. Or build partnerships to work together.

If you are going to start a new organisation it is important to know what will make it unique and clearly articulate this. What will you do differently? Which gap are you filling? Why is no one working on this issue within your geographical focus? What is innovative? If there are other organisations that are doing broadly similar work to the project(s) that you have in mind, think carefully about whether you would best meet your aims by starting a new organisation, or whether you can best do so by working for, or in partnership with, an existing organisation.

Given the amount of coordination and administrative work that needs to be invested in NGO management (and the corresponding costs to this), there is a need to avoid duplication, and there are many advocating for organisations working in a similar area to merge to work together more effectively, having their separate programmes under the same umbrella NGO. If you are to be an international NGO (based in a country other than the one you are operating in) you will definitely need to have a local NGO or CBO partner in order to help you to implement the work.

> Malawi has a huge shortage of health professionals and the statistics on numbers of doctors is dire. There are huge efforts being invested to train up new doctors, but when I was there on elective educational placement, I found that many of the Malawian medical students were having to drop out of their studies, due to lack of money to pay tuition fees or because they needed an income to support their families. I felt a lot could be done to support these bright young medical students to continue with their studies. This seemed like a great opportunity to engage colleagues back in the UK to provide direct sponsorship to medical students from poorer backgrounds. That was the birth of Medic-to-Medic.
>
> As it would be a small organisation, I decided to try to set it up in collaboration with another organisation with a similar mandate. As a new starter, this was a lower risk strategy compared to starting up a new organisation from scratch on your own. You're working with people who have made it happen and who know the ropes, which is very instructive if it's all new to you. In addition, this approach can ensure that you are filling a niche, rather than duplicating other people's work. Other pros mean that you don't have to register with the Charity Commission (in the UK), which can be a lengthy process, nor file yearly reports to them. On the downside, as an 'add–on' to the organisation's core activity, you may have to work harder to justify the need for your new programme. Seven years on, we had grown sufficiently to bring all our programmes together as an independent charity, Health Workers for All.
>
> *Kate Mandeville, Medic-to-Medic/Health Workers for All*

Funding

The survival and growth of an organisation depends on whether it has the finances to implement its activities and projects. The most efficient charitable organisations channel the largest percentage of funds into running the project, minimising administration costs. Nevertheless, you are likely to require some start-up capital, have running and administration costs (including website hosting, communications costs, staff salaries for example), and of course need funding for your main purpose: project implementation. You need to establish a budget for your organisation and think about where this money will come from. Will there be any paid staff? What is the minimum funding required to implement your work? Are you eligible for any grants or funding? Will you be registered as an NGO or have a social enterprise model? These are just some of the questions you have to ask yourself. Develop a budget that includes running costs, capital costs and project implementation costs; factor in evaluation too. As a general rule within the industry, running costs should be no more than 10 per cent of a total organisation's budget, with evaluation costs often accounting for another 10 per cent. NGOs also need to be legally registered (a lengthy bureaucratic process in most countries), and provide yearly financial and activity reports to stay registered. They will often get tax breaks but in many countries need to be governed by a board of trustees.

SOCIAL ENTERPRISE?

A social enterprise is a business driven by a social or environmental purpose. As with all businesses, they compete to deliver goods and services but they are usually started by a person or group of people with a particular passion or sense of purpose, who didn't set out to create a business at all. The difference between a social enterprise and a normal business is that the social purpose is at the very heart of what they do, and the profits they make are reinvested towards achieving that purpose, rather than it being driven by the desire to maximise profits for shareholders and owners.[1] Social enterprises may take on different legal structures, depending on the country where they are registered. The decision to go down one route or another will depend on the type of organisation you are trying to set up, and what the purpose is. Do you have a marketable product that people may want to buy from you? In some countries charities are not allowed to 'sell' products, and to make an income by any other means than donations or grants, where they are accountable to their donors. Some well-known *examples* of social enterprises include *The Big Issue* and the fair-trade chocolate company *Divine Chocolate*.

I co-founded Harambee Schools Kenya (HSK) after a gap year teaching in a rural secondary school. When I returned to university I wanted to do something other than sitting in the library or pub. We also knew that people in our network wanted to help and having a legally registered charity enabled us to be more efficient (especially using the Gift Aid mechanism for reclaiming tax). Over the years we have received most of our funds from grant-giving trusts, made possible by the expertise and connections of one of our trustees. It was part time, managed in spare time with the broad support of trustees and other people.

In 2007, after a number of years in corporate jobs we co-founded Skills Venture (SV). Through HSK we received several emails from people who wanted to come out to Kenya for a few weeks to help us build schools. We turned them down, on the basis that we were hiring Kenyan locals to do this, and didn't want to put them out of work. But we realised that there was huge demand from people in the UK to do something useful overseas, and that many of them had years of professional experience in areas such as marketing or HR which were useful skills to transfer in Kenya. SV was founded as a social enterprise, generating all of its own income from charging volunteers to organise their placements, rather than being dependent on external funding from donors. We chose to structure it as a Company Limited by Shares, because we wanted to have the option of selling some of our equity to investors to raise capital if we needed it.

We had spent eight months developing the concept and writing a business plan before we handed in our resignation letters, but we were still a long way from generating enough income to cover anything like two reasonable salaries, and we were too naive and optimistic to realise this at the time. Our savings only lasted us for a few months so we had to take on part-time consultancy jobs to cover costs, which meant less time to invest in the business. In reality this needed seven days per week for the first couple of years to set it up, so it inevitably suffered. The second challenge was that we didn't benefit from the support and advice of the trustees like HSK did. The main lesson that I learned from this experience was that when selling to clients and donors it is very important to distinguish enthusiasm from a willingness to part with money and to buy your service.

Overall, I think that being a founder of HSK and SV has helped my career prospects. Most employers look at this evidence of entrepreneurialism as a positive trait, and I have no shortage of examples to show what I have learned along the way in terms of skills and experience.

Will Snell (UK)

Implementation

Starting up and growing a successful organisation is an extremely time-consuming endeavour. Do you have the time and skills to see this through? You need to make a realistic assessment of the size of the organisation, how you expect it to grow, and what time and knowledge investment are necessary to achieve your organisation's goals and objectives. Very few well-known NGOs today made it big as a part-time endeavour. Can you commit to the necessary investment of time and money (you might not have a salary for several months)?

I have seen the work of many small NGOs, and very few of the ones that are started as an individual venture or to fulfil someone's own employment needs are successful. On the flip side, those that have a strong vision and a mission, and have been successful in engaging others with the right skills to share their vision and join the journey, have.

Anonymous

Think of all the things that need to be put in place in order to start and establish an organisation: you will need not only to raise funds and put systems in place for managing these funds and other activities, but also to manage the project and implement the work, liaise with partners and other organisations, create effective communications for your target audiences (website, blog, annual review), submit reports to your donors, trustees and the charity commission – all this in addition to your main purpose: the programmes you are implementing, support you are providing or your advocacy efforts. It's a great learning experience, but it will be difficult to do it on your own. You will need to bring other people along for the ride with you.

TOP TIPS

- Start with the beneficiaries – the people who need your help – rather than with a proposed solution. Be careful of imposing your great ideas on people who might have other, more pressing, needs and interests.
- Be prepared to spend years getting things up and running. You should also expect some initial hardship, setbacks, and financial constraints – there's no rule book, no silver bullets, just consistent very hard work.
- Allow for your ideas to change over time. Learn as you do; don't try to learn first and then implement a fully developed idea.
- Don't be surprised if people don't share your passion or understand your cause. You may need to be the one who spearheads change; others will follow eventually. But just because *you* see something as being important and worthwhile doesn't mean that others will too.
- Do your research, as thoroughly as possible, to work out which organisations are already working in the 'space' that you are interested in. If no one appears to be in that space, make sure that you haven't missed anyone. If you haven't, think seriously about why no one is doing what you are thinking of doing. Maybe there is a good reason for this! How do you view them . . . as collaborators or competitors?
- Try to identify potential partners, suppliers, colleagues, and most importantly funders or clients, before you make the leap. But don't expect to have them lined up and ready to go – you will need to make that leap first, and hopefully they will follow!
- Try to think about your long-term career plans. Do you want to be running your organisation until you retire? If not, at what point are you going to move on, and what do you need to do now in order to make this as easy as possible? How will you do this, and what kind of person or organisation might you hand it on to? What will you do with your life once you have left, and what skills and experience do you need to gain through your time running the organisation to make this transition as easy as possible?

Engagement with the government is crucial in humanitarian/development work and if you are working outside your own country, it is likely that you will need to be legally registered in the place you are working as well.

Think carefully about who to bring on board. Do you need like-minded people, or those that demonstrate a diverse range of skills: the visionary, the executor, the promoter, the advocate and the academic? Bring together a team of people so that together you can make a strong idea a reality. Getting others to share their skills and expertise with you as you build your organisation can sometimes be easier than obtaining money – and can often be more effective. Seek out good team members to join you – in the knowledge that you will need them to work alongside you in a committed fashion. With this in mind, be prepared to delegate ownership and responsibilities to allow others to grow alongside you. Make sure that together you have the necessary time to invest in the organisation and make it grow.

Evaluation

Being able to accurately measure and report to your donors or stakeholders on the impact of your work will be crucial to develop a track record and help you to grow the organisation. Think early on about what the success indicators will be, how you will measure these. In order to do this you will need to have baseline data – the situation at the start, to allow for comparison later on. Set targets, in numbers, to be able to monitor progress as the organisation sets off. But be realistic. One of the main reasons why many development projects seem to fail is that they set incredibly high and unrealistic goals. So make sure you don't set the bar too high.

Learning and sustainability

All organisations need to consider sustainability and think strategically about how they will continue to fund their programmes beyond the current round of grants. Any organisation becomes vulnerable if it relies on the enthusiasm and motivation of just one or two people. Some people like to stick to things, see them through until the end, while others prefer to be the catalyst, the founding energy, and then hand things over to others to run.

As founder it is important to know how long you can commit for, and if it is to continue after the end of your time, who will take over. What is the future of the organisation? What is your exit strategy? How will you engage the continued support of the organisation? If you are providing services, you should not be absolving the government from their responsibilities but engaging them as partners to potentially, eventually fund or take over the work.

IN FOCUS: AFRIKIDS

While travelling in Ghana during a gap year, I spent some time volunteering at a babies' home run by a local nun who rescued 'spirit children' (ostracised by their families due to harm caused to the mother in childbirth, or with physical defects). I was blown away by the success she was having in reintegrating them into their homes and communities as well as tackling such entrenched beliefs with so few resources. I decided to support her through direct fundraising during my university years.

In 2002 I established AfriKids to formalise my support and started working on it full time. We grew exponentially and by 2005 I had built a strong team around me both in Ghana and the UK. Our project work is focused on direct support for children and ranges from initiatives that resettle children from work in gold mines and life on the streets, to holistic community education and skills programmes and empowering mothers through microfinance. We focus on a very specific region, which has allowed us to work with staff who understand and are deeply integrated into their beneficiary communities.

We started investing in social enterprises and set a firm goal for local sustainability by 2018. The social enterprises we establish will not only provide independent financial support for the future work of AfriKids Ghana, but will also provide employment and skills opportunities to the children supported by our programmes, as well as providing a significant boost to the local economy and wider prospects for the region. The aim is to pursue a new, effective, transparent and holistic approach to sustainability, which I feel is lacking across many NGOs.

Starting an NGO is a great leap into the unknown but ultimately much of the work requires common sense, good judgement and a focused goal. The first months – indeed years – of establishing an NGO can be extremely tough in terms of generating enough funding and momentum to see you through, especially when you need to continue to support yourself financially. I had to work hard to engage donors to share my vision, with little substance to show at the beginning except my track record of voluntary work with good (but limited) results. The money for wages was not always there – that's tough but something you need to be prepared for.

I realised that our success was dependent on engaging the right people, so I got them involved without hesitation. We have also developed a partnership model with our supporters – keeping them intimately involved with our work. It's a very intensive way to work, but after many years these efforts continue to pay dividends and to shape the way we interact with our supporters. Engaging outsiders with specific expertise has also become more important. Two examples are our Hospital and Eco Lodge. I found it easier than I imagined engaging external voluntary expertise. We've found that our projects provide a great opportunity for professionals across all walks of life to contribute to exciting social programmes using their skills, and in ways they find engaging, meaningful and satisfying. So there really is a two-way partnership that we can offer – and that works very well in practice.

Our annual turnover is now £2 million. Most of it comes from companies and foundations (we were Deutsche Bank Charity of the Year in 2010 and raised more than any other charity has there in a single year – over £1 million), individuals, events and statutory grants (DFID, Comic Relief, UN, ILO and the Big Lottery Fund).

Being the founder of a successful organisation brings recognition and prestige. Our directors are regular speakers at corporate, social-enterprise and third-sector events of many different varieties. We've also found that other charitable organisations increasingly come to us to obtain advice on a range of issues, from emulating our delivery or fundraising success, to the challenges of running charitable organisations, and as a result we established a small consultancy, AfriKids Squared. In 2013 AfriKids Squared advised 50 organisations in 20 countries.

Georgie Fienberg, Founder of Afrikids. www.afrikids.org

Putting your idea in motion

If what you have read above has given you the conviction to continue with your organisation, then it's time to take it from conception to reality. Here is some useful reading to help you plan further:

Robert Ashton (2010) How to be a Social Entrepreneur: Make Money and Change the World, Chichester: Capston Publishing.

David Bornstein and Susan Davis (2010) *Social Entrepreneurship: What Everyone Needs to Know*, Oxford: Oxford University Press.

David Marfleet (2011*) Grow Your Own Charity: Down-to-earth Essentials for Managing Good Causes*, Kibworth: Troubador Publishing.

Rorey Ridley-Duff and Mike Bull (2011) *Understanding Social Enterprise: Theory and Practice*, London: Sage.

Note

1   www.socialenterprise.org.uk/data/files/publications/Social_Enterprise_Explained_-_May_2011.pdf (accessed 30 January 2015).

12

MOVING ON

Humanitarian work, in highly intense and stressful environments, is not always conducive to a stable life, let alone a family. What may have been attractive and exciting about the sector and work at the start of your career in the field – and satisfied your sense of adventure – may leave you pining for quite the opposite a decade in – or earlier. Even those working in the more stable environments, in reconstruction and development efforts, based in a country other than their own, may reach a stage where they want to move on, relocating back to their home country. Some may get disillusioned and prefer to leave the sector altogether, or go on to a related field. This short chapter shares some of the experiences of others moving on.

Returning *home*

> I have met quite a few people who love the idea of development work when they are young, coming out of university, but as they get a bit older and start thinking about serious relationships and family it gets more difficult and people start to reconsider whether it's actually the career for them.
>
> *Career advisor*

Not all development jobs will be based outside your home country, but when they are, and you have spent a long time in different cultures, outside your roots, you may reach a point when you want to return *home*. If you choose to live in a place where there is a lot of international activity, you will find more opportunities. With field experience, many people are an asset back at headquarters, enabling them to build on this to support country teams, in management work or HR for example. But these positions are relatively few and of course not suitable for everyone. Those who thrive in an operational role on the ground, especially in the high stress of emergency contexts, may be less suited to the slower and more stable pace of HQ.

> I went to Africa with my wife. Our children were made and raised in Mali and Tanzania. When my eldest daughter was 12 years old, we decided to move back to the Netherlands as we wanted our children to have their own spot on the globe and build their cultural identity and social networks from there. I am now based in the Netherlands but spend six months a year on short-term assignment in Africa, managing or contributing to large projects and conducting short assignments (project design, training and coaching, monitoring and evaluation). I also teach on a number of courses back in the Netherlands.
>
> *Ted (Netherlands)*

There are several areas of work that are at the edge or periphery of development and humanitarian assistance. One of these is human resources, for example, as getting the right people in the right jobs is better done by those who have an understanding of the sector, the conditions and already have a good network of contacts in the field. Proposal writing, fundraising, advocacy work or management are other possible roles back at HQ. Many people may go on to research or writing jobs too.

Consultancy work (see Chapter 10), or short-term assignments are a good option for getting back to a good work–life balance. This will enable you to draw on your field experience and apply it to a range of contexts and organisations, while maintaining a link to their field. By taking on a few assignments per year, you can essentially live wherever you call home the rest of the time.

Challenges returning home

Integrating back home can be as difficult as moving to a new country, as things are likely to have changed significantly, especially if you have been away for some time. Many of your old friends will not have the international outlook that you have gained, or even begin to understand your experiences.

If you have worked in humanitarian contexts and high stress environments witnessing distressing things, the return home is difficult. Some organisations will have a system in place, such as a network of returned professionals or volunteers, enabling you to link up and get support from each other. Others might offer psychological support to help you talk through or make sense of your experiences. Ask your organisation what might be available.

If you are returning with the additional burden of trying to find a job, settling in, making new friends and so on, it can be a stressful time. Make sure you ask for support from those around you or even from your organisation or other professional support. Above all give yourself time to settle back in and manage your finances well.

Moving to a different sector

> After several years working in the international development sector, I started to become a bit sceptical about the whole development concept, and the idea of trying to *develop* countries in Africa. As I tried to make the transition to jobs back in Ireland – I was particularly interested in working with marginalised groups such as travellers and asylum seekers – the fact that I didn't have much professional experience in Ireland was a disadvantage. After a time, I stopped focusing on applying for positions that were advertised in newspapers and joined a community-based organisation as a volunteer. The work enabled me to make friends, learn more about my new context and to build up a network. A year later I managed to get a paid position with a very good organisation.
>
> *Sive (Ireland)*

It's not uncommon to find that those who were idealistic and ambitious at the outset become disillusioned after direct experience of working in the sector. Couple this with the frequent lack of clear career path and short-term nature of the work, and some might chose to go down a more conventional route, retraining if necessary.

> I worked in South Africa with women with HIV and then back in London as a programme manager for an international NGO. While I enjoyed the work, when I was 32 I experienced an overwhelming sense of not having a profession. All my friends were now accountants, doctors, management consultants or journalists moving up their respective career ladders and I couldn't see a clear path for me. So I managed to get a training contract with a law firm, and made the transition to becoming a lawyer. In future, I may be in a good position to combine law and development, but for now I am happy where I am.
>
> *Maaike (Netherlands)*

But frequently those working in the sector end up with more questions than answers, so research, writing and academia are other common routes to go down.

> I moved from humanitarian work into academia because I was interested in understanding the broader political, social and economic systems that were creating human suffering. Disasters, conflicts and crises are expressions of and the result of how we organise our society. Too often aid workers are remarkably blind to the context in which they work. In my own work I made many mistakes that derived from my basic ignorance of the situation around me. I wanted to go to graduate school to study the political reasons for and implications of the mistakes I made. This led me to do my Master's and PhD on the politics of foreign aid. Having some experience of 'field realities' has been very useful to my academic career – it has grounded my more theoretical work. I think that writing reports for NGOs also taught me to write with clarity. I still stay in

touch with the humanitarian sector, through occasional consultancies, advising NGOs and discussions with old friends – this helps to keep me current.

Matthew (USA)

Transferable skills

Whoever your potential next employer, and whether staying in the same field or moving to a completely new one, don't ignore the skills and experiences you will have built up. You will have likely gained many transferable skills, but often only those recruiting managers who themselves have overseas experience will understand the true value of it. Some that you may like to highlight include:

- managing complex programmes across cross-cultural environments;
- gaining good writing skills, budget skills, working with donors;
- navigating a handful of complex situations that you have had to deal with.

When our son was born my partner wanted me at home more often. But not wanting to let go of my career – nor going back to live in the UK – we decided to set up a sustainable tourism project in north Africa. My cross-cultural skills, managing budgets and complex operations served me well as we worked with local craft people to renovate our 300-year-old house. I still stay on emergency rosters and do a couple of assignments per year having most recently spent three months in Haiti.

Anonymous (UK)

PLATE 7 'I have been working on the Control Arms campaign for the last ten years. This picture shows mock tombstones that were placed along the East River by the Control Arms Coalition to coincide with a diplomatic conference on the Arms Trade Treaty in New York. There isn't one usual day working in campaigning. One day I might be in and out of the UN – talking to diplomats, speaking to the media and meeting colleagues from other campaigning organisations to discuss and agree strategies. Another I might be at the forefront of a public campaign to draw in media attention. Successes take time but at last the world has an Arms Trade Treaty – a huge campaign win.' Anna McDonald.

Source: Andrew Kelly

PART 4

Areas of speciality

What follows are 54 different thematic areas, job functions or areas of expertise within development and humanitarian assistance sectors, listed in alphabetical order. This section highlights the huge scope of work within the fields of development or humanitarian assistance. It will help you to understand the areas of specialism that appeal to you the most, and what transferable skills you may have within these fields, and thus help you to further direct and plan your career.

One of the most interesting aspects of this section is the individual people's stories: their career paths, what motivates them in their work, and what their daily focus is. Some are career changers, others sector changers, some ended up in their field quite by accident, for others it was a long and deliberate process. While people at all stages of their career are featured, the tendency has been to select people in early to mid career so that their entry route is not too dissimilar from that of someone wanting to enter the sector today. In brackets is the person's country of origin, followed by up to five developing countries where they have worked. Many have of course worked in a much greater number.

There are four important things to highlight about each of the specialities listed:

- First, the list of 54 is by no means exhaustive. There are many other areas of focus that, due to both time and word constraints, could not be included. Featured are some of the most common fields of work and also some of the more unusual emerging areas. The field of international development and humanitarian assistance is in constant flux with changing demands. Areas that may be in fashion today, may be out of fashion tomorrow. If this list is redone in ten years' time, some very different areas and trends may appear.
- Second, development work is, by nature, both broad and cross-cutting. It is rare to find someone who works exclusively in a single one of these fields in isolation. In reality, most people's work will touch on several of these areas

at any one time. For example, a person working on an agricultural programme, focusing on value chains, may be doing a programme management role. A large component of their job might also be grant writing, and perhaps a bit of advocacy as well. Of course they also have to be aware of gender issues and be working to enhance people's livelihoods and food security: the ideal T-shaped person (see Chapter 9).

- Third, each of these individual *specialisations* is in itself broad in scope, much more than can be adequately covered in this introduction. Inevitably, generalisations have had to be made and the entire scope cannot be addressed. The case studies provide a snapshot of one person's work in that field, and are not meant to be comprehensive or even representative of all work within this field. If you are interested in more depth, the *Find out more* resources are good places to start.
- Fourth, some specialities are primarily relevant to development work, others to the humanitarian context, but the majority can be applied to both development and humanitarian contexts.

The icons used throughout this section have been produced by the United Nations Office for the Coordination of Humanitarian Affairs (OCHA) or sourced from The Nounproject http://thenounproject.com

ADVOCACY AND CAMPAIGNING

Advocacy or campaigning (you might also call it activism or influencing) involves creating change and improving people's lives. It is the deliberate process of influencing those in power, through the creation, reform, implementation and enforcement of laws that protect and promote the interests of those you are speaking out for – be it those with no voice or the under-represented. The work goes hand in hand with policy analysis and reform. Whether it is to protect a remote rural community in danger, or mobilise donors to end a famine, you will need to passionately argue for the issues that you are advocating (and, ideally, believe in them), and be able to put your point across convincingly to bring other stakeholders along on the journey. Any successful advocacy campaign involves strategic communications and speaking the same language as those you are trying to influence.

The work: You will be targeting specific audiences to achieve your desired outcomes. An effective advocacy campaign will be multidimensional and utilise a range of approaches. Research is a significant part of the role, as you first need to have the facts to hand in order to convince others. Writing articles and publishing findings may be one way to mobilise the public. Bringing other people along with you is important, so developing effective relationships with other civil society organisations, including through formal networks and coalitions, is also a key aspect, as is forging new and innovative partnerships. Fundraising is often linked to some campaigns, mobilising funds to make things happen. Outcomes can be slow, but the impact is the real change created by a campaign – the difference it makes to people's lives. But patience is key, as this sometimes take years.

The employers: Human rights organisations, lobby groups, pressure groups and rights-based organisations such as Amnesty International, CARE International, Save the

Children also have a strong advocacy component. An increasing number of donors, NGOs and funds also emphasise advocacy, which they see as a way to achieve widespread, sustainable change.

Breaking in: Mostly it is learning by doing although knowledge of communications can be a plus, as is a background in research or advertising. Having a strong passion, stand or sense of injustice is a good starting point. This involves an in-depth understanding of the issues, and first-hand contact with the people/issues you are speaking for. The ability to put forward coherent arguments that will be considered by others in an opposing field of thought is another. If you are dealing with legal reform a background in law or in public policy will put you at an advantage but it is not a prerequisite.

Potential job titles: Advocacy Manager; Policy Analyst; Campaigner

Find out more: Many job websites list advocacy and campaign jobs.

Ellie Levenson (2011) *50 Campaigns to Shout About*, Oxford: Oneworld Publications.
VSO Participatory Advocacy Toolkit, www.vsointernational.org/Images/advocacy-toolkit_tcm76-25498.pdf

See also: Fundraising; Communications; Journalism

The injustice I lived through during the genocide in Rwanda in 1994 is the driving force behind my work. I am lucky enough to have had access to a good education and have the confidence to speak to people in high places. I studied law, and graduated six years ago. My first job out of university was with a legal rights group, campaigning for the rights of *génocidaires* in prison. Personally I found it difficult as I was working for the people who may have been involved in the genocide, so I didn't fully believe in my work.

I then started working as a project officer at Survivors Fund (SURF) and was then fortunate enough to be selected on to the International Research and Exchanges Board (IREX) community solutions programme. This is a fellowship funded by the US state department for young professionals from developing countries to spend six months with an organisation in the US. Of course I chose the advocacy strand. My time there gave me a good insight into a professional advocacy organisation and the state of advocacy in the US.

I have rejoined SURF since returning home and am heading up their legal advocacy project. Here I am fighting for the rights of genocide survivors. This is a cause I can fully stand for, am passionate about and feel that what I am doing is right. Genocide survivors are marginalised, stigmatised and disadvantaged. People perceive them as searching for pity. They see themselves with guilt: why did we survive when so many others did not? Survivors have lived through, and continue to live through, injustices. The main thing that we are doing is challenging the laws and policies that hinder genocide survivors' access to effective reparation. SURF is working with human rights organisation REDRESS, in collaboration with Rwandan civil society, to examine survivors' experiences in seeking reparation. The work has been shaped by the closure of *Gacaca* (village-based trials) as well as the UN International Criminal Tribunal for Rwanda in 2012, and serious concerns among survivors that with the closure of both mechanisms, avenues for reparations would be closed, too, and their right to reparation would never be met.

A typical day for me might be arranging lobbying meetings with key people surrounding decision makers, defending a response to a bill in parliament and organising advocacy workshops for survivor-led organisations. But any information has to be backed by evidence, so equally I have to do research and build up case studies of the people affected, so I also spend a lot of time speaking to and interviewing survivors, especially widows. I recently also worked on a submission to parliament. It is exciting work, but slightly insecure, as sometimes we can be perceived as a threat, challenging local authority or taking a stand against government who think it is time to 'turn the page'.

To work in this field be passionate about and believe in the cause you are working towards. You need to be convincing and be able to influence, although this doesn't mean being loud or outspoken.

Albert Gasake (Rwanda)

CAMP MANAGEMENT

With millions of refugees worldwide, and an escalating number of internally displaced people (IDPs), camp management is a crucial part of any effective humanitarian relief response. Those who have been forced to flee their homes due to conflict and natural disaster, leaving relatives and belongings behind, often have little choice but to collect in camps for assistance, safety and security. Camps are usually created to help the concentration of people fulfil their basic needs of shelter, food and hygiene. This is complex and requires careful management. Local government and humanitarian organisations come onto the scene to help set up and manage the camps, some of which may only be needed for a few months – although in reality many camps last for several years and sometimes even for decades. Regardless of lifespan, camps can only offer temporary assistance, and the role of the camp manager is to raise the standard of living in camps and camp-like settings, and to uphold the rights of camp residents. The importance of camp management as a vital component of humanitarian relief has been recognised by the Global Camp Coordination Camp Management (CCCM) Cluster.

The work: Camp management focuses on developing effective coordination mechanisms that support the provision of assistance and protection. Camps do not provide durable solutions to displacement, but the aim of the camp manager is to ensure coordination between all stakeholders, facilitate distribution of material assistance, identify gaps in services, avoid duplications and ensure that the fundamental human right to life with dignity is upheld for the camp community. You will be involved in coordination work with stakeholders, information management and registration, the local environment including housing, sanitation, livelihoods, livestock and farming as well as site planning, protection and eventual closure.

The employers: Camp coordination is often performed by a local or international NGO/agency such as UNHCR, IOM, IRC, etc. A diverse range of organisations also play a role in providing services in camps, such as WASH, psychosocial services, food assistance, education and providing technical assistance including Catholic Relief Services (CRS), Danish Refugee Council (DRC), Norwegian Refugee Council (NRC), Shelter Centre, OCHA and CARE, to name but a few.

Breaking in: Gain experience of working in camps, managing specific programmes within them. Then take a training course on camp management; these are widely available.

Job titles: Camp Manager; Camp Coordinator

Find out more: The NRC offers a toolkit, as well as training and coaching. The CCCM produces an annual newsletter on updates in the field.

The Emergency Information Management Toolkit: data.unhcr.org/imtoolkit

See also: Shelter; WASH; Humanitarian response management

After obtaining a Bachelor's degree in Anthropology with a minor in African Studies, I spent some time in Africa studying and volunteering to get a better idea of the type of work I wanted to do in the long term. I then completed a Master's degree in Public Health and Social Work at Columbia University, specialising on issues of forced migration and health. I focused a great deal of my research on the IDPs in the eastern region of the Democratic Republic of Congo (DRC) and found that this area of the world was extremely interesting and complex.

I knew the key to securing employment after graduation was to expose myself as much as possible to the field of humanitarian aid work: hands-on field experience was essential. While focusing on my studies, I also sought out internships and was fortunate to find one with a large humanitarian INGO whose HQ are in New York, where I was studying. I began first as an intern and was eventually offered part-time employment. I travelled to Africa (Uganda and DRC) with this INGO before graduating from university. On my first trip, I assisted in organising a conference and on my second I actually gained experience by briefly working in a refugee camp hosting Angolan refugees in SW DRC. During this time, I also began to study both French and Swahili outside my regular coursework to strengthen my CV and spark a potential employer's interest.

Just before graduating I began my job search and was offered a position with a small INGO in eastern DRC. I spent four years working in this area, first managing health, nutrition, and gender-based violence (GBV) programmes for IDPs and was later promoted to become the Deputy Country Director, managing larger operations covering IDPs and host populations through health, HIV and AIDS, nutrition, food security, GBV, WASH and IDP camp management. After briefly returning to the US to conduct some research, I accepted a position overseeing all programmes for several

Congolese refugee camps in Rwanda. The context here was different, as more funding meant well organised camps and some of the refugees had been displaced for nearly 20 years: most of the children had never been to the DRC.

I now work for International Rescue Committee (IRC) in Central African Republic (CAR). We work with IDPs that have fled recent and previous fighting. A lot of my time is spent in meetings but I also go on occasional field visits. Tomorrow we will drive 12 hours on a dirt road up country to visit health posts, markets we're building, and also visit communities as well as negotiate with local rebels who have just taken over the area, to ensure we get humanitarian access.

I love the work that I do; I like to see the impact I make on individual lives. On the flipside it is emotionally stressful and can be exhausting working in a highly insecure area with repeated outbreaks of fighting and violence. In these settings, you are never 'off' and 70- or 80-hour weeks are the norm when responding to a new humanitarian crisis. This work is not for everyone, but for me it is profoundly rewarding and I continue to love it despite the stress and hard work it demands.

Sarah Terlouw (USA), Central African Republic,
Rwanda, DRC, Uganda

CAPACITY BUILDING

This has become a widely used term within development and most projects, whatever their remit, will aim to have an element of capacity building. There has been some controversy over what this slightly vague term means. Capacity building is linked to the concept of sustainability – the ability to endure over time – and is generally accepted to be the process of building up the knowledge and skills of local staff, or even whole communities, enabling them to continue or lead a process of change. This can be done through training, formal education, mentoring and support. By building local knowledge – the rhetoric goes – external aid agencies and experts can eventually pull out, thus reducing donor dependency. Building local capacity is a specific form of assistance within many funding streams as an essential component of development. The term gained prominence in the 1990s due to the growing realisation that technical solutions and funding alone were not sufficient to address most development challenges.

The work: Capacity building efforts might be focused on government or civil society and many jobs, especially those based in the field, will incorporate an element of capacity building and knowledge transfer especially for international recruits and experts. You might be training others to ultimately be able to do the job you are doing, or to do their own job more effectively. The process usually has several stages and will start by identifying the barriers to local implementation of a project. A plan will then be developed to address these capacity shortfalls. The capacity development response might include systems, leadership, knowledge and accountability. You will then be involved in overseeing or implementing the response and finally evaluation. Training local staff and preparing them to manage resources effectively often requires a huge investment – and a system of checks and balances – so evaluation is a key component of the role.

The employers: Most organisations will work, in one form or another, in capacity building. Some organisations will also employ 'capacity building experts' who solely focus on this, but more often than not, capacity building is a skill expected of most experts positions in whatever field they specialise.

Breaking in: Some short courses exist, such as training of the trainers (ToT) courses, but mostly it is about learning good management techniques on the job and being a good coach and trainer. A Master's in Adult Education can be a plus. Facilitation and interpersonal skills are also essential. Mostly you will be building capacity in your technical field, whether it is neurology, project management or advocacy, so knowledge of this area is the most important.

Job titles: Capacity Building Advisor; Coordinator; Specialist

Find out more:

Deborah Eade (1997) *Capacity-Building: An Approach to People-Centred Development*, London: Oxfam Professional.
www.capacity.org a gateway to capacity development

I joined Crown Agents on their training scheme (now discontinued) immediately after I graduated from my BSc in International Disaster Engineering and Management at Coventry University (UK).

I manage a project on the modernisation of Zambia's Medical Stores Limited (MSL). In 2004 the Zambian government contracted Crown Agents to manage this government-owned entity under a management contract arrangement, which is responsible for supplying medical commodities to hospitals and health centres throughout the country. The efficient running of this organisation is vital to ensure efficient delivery of health services to Zambia's 13 million population. But when the project started we found a physical infrastructure sorely in need of investment, inefficient operations, expensive distribution arrangements and a dispirited workforce. All of this was costing the Zambian people a lot of money for an inadequate service. We set about to change this.

We initially put in a senior management team of Crown Agents staff to manage the organisation and the modernisation programme, but the aim has always been to build up the capacity of the MSL staff and transfer the senior management positions over to them, so that they can eventually run it and take the work over themselves. Overseeing this capacity building process is one part of my role.

To give an example of this process, the senior logistics manager, a MSL employee, was identified to become the director of logistics. But he needed to gain additional skills and training in order to step up to this role. We sat down with the team, established what level of knowledge and skills he had to gain before taking on this role, and identified where these gaps were. We then developed a plan for him to gain these additional skills, and monitored progress carefully. One element of his training involved him undertaking an online Master's course in Logistics. It was a very formal process and once we assessed that he had gained the relevant skills, we made a recommendation to the board for him to take over this role from the Crown Agents Director.

In a development context we use the term capacity building but I also do similar work with my team back in the UK, yet here we call it training and professional development. What I enjoy the most about the job is that by building up this essential service we are improving drug availability for essential medicines and for HIV and AIDS, Malaria, TB and other diseases.

It is making a difference to people's lives and that is my main motivation. I also get to travel to remote parts of the country, to visit the health facilities and district stores. There aren't many lows as I really enjoy my job. I guess one is having to write proposals to bid for work. And the speed that things happen on the ground is frustratingly slow sometimes.

Sian Rogers (UK) Zambia, Zimbabwe, Botswana, Nigeria, Sudan

COMMUNICATIONS

Communication jobs are dynamic, intense and full of creativity. Successful communication in the development and humanitarian sector demands a variety of communication skills and specialities. From branding, to corporate, internal, external and specifically 'development' communication, you can choose to either specialise or become an all-rounder. Members of a communications team in this digital age actively search for new and innovative ways to get the message out, through social media and mobile technologies, to reach targeted audiences.

The work: There are three main areas of specialisation but most jobs will have you involved in at least two of these. Working in *internal communications* you will focus on enhancing organisational structures and productivity within teams. You might be streamlining presentations, developing internal newsletters and managing the intranet. Working in *corporate communications* you will be outwards facing, focusing on branding and PR for fundraising or advocacy. Your target audiences will be governments, other large organisations and donors. You might be writing press releases, developing success stories, creating content for the website, managing the filming of a video about a project and assisting the management with branding of the organisation – developing logos and graphic design manuals. And finally, Communication for Development (C4D) aims to bring about positive social change, focusing on practical communication with project beneficiaries – these may be farmers, children, women of marginalised communities and so forth. In this case you might be doing research, developing strategies and identifying the right medium and messages to reach out to your target groups. You will need to know your audience very well and develop long-lasting relationships with media in the region.

The employers: All organisations that have projects running with an external audience and donors need assistance in developing and fine-tuning communications to ensure a successful project. Most large organisations, be it an NGO, think tank or donor who maintains a public profile, will need someone in charge of their communications. Smaller organisations are likely to combine communication with other specialisations, such as fundraising, training, and knowledge and information sharing.

Breaking in: A background in communications or media, be it through formal studies or experience, is a good starting point. A good basis of development theory will help you understand the context you work in. You will need to be a good strategic thinker. Strong writing and visual skills are a plus, as long as you are capable of implementing them as part of a bigger strategy.

Job titles: Communications Officer; Communications Specialist; Marketing Manager; Press Officer

Find out more:

Nigel Hollis (2010) *The Global Brand: How to Create and Develop Lasting Brand Value in the World Market*, Basingstoke: Palgrave Macmillan.
Thomas McPhail (2009) *Development Communication: Reframing the Role of the Media*, Chichester: John Wiley & Sons.
Linje Manyozo (2012) *Media, Communication and Development: Three Approaches*, London: Sage Publications.

See also: Journalism; Advocacy

The challenge of really understanding individuals and groups, their needs and problems, and helping them to find solutions was what made me want to work in this field. Communication has the power to give people a voice, creating awareness, and understanding. Through communication people and projects can become more interactive, which leads to ownership and empowered audiences.

My initial background lies in advertising. I went to art school and worked on campaigns for big brands. But I wanted to work in development so I pursued a Master's degree in Global Communications at the American University of Paris where I specialised in 'development communications'. The combination of theory and practice in India and Morocco made me realise that this was the career for me. I then worked as Operations and Communications Manager for a grassroots NGO in northern India and I am now Development Communication Specialist for International Fertilizier Development Center (IFDC) in The Great Lakes Region (Burundi, Rwanda and the DRC).

These days being home is the exception. I am spending at least half of my time in the field and in the region. I supervise development communications for three different countries, support content development, meet with service providers, make sure our strategy gets implemented, while keeping an eye on the budget. Right now we are working on setting up interactive radio shows. We are researching tools and methods for generating and managing feedback, using old media such as question-boxes at radio stations, but also looking into the possibilities of new media such as mobile technology and text messaging. We are no longer just sending out one-way messages but listening to what our target group has to say and responding to that. It might sound basic to many, but it is still quite a new approach in many developing countries. Together with other media tools such as mobile cinema, debates, theatre, knowledge centres, events and many more, we strive to increase interaction and thereby ownership of the project by our beneficiaries.

There are a couple of trends that are worth watching. First is the shift in focus from a top-down approach (where organisations and governments feel they know best) to a more bottom-up approach, which gives people the opportunity to have a voice within projects. This is the so-called 'Development Communication'. Second, rapid technological developments have provided us with new exciting tools such as mobile phones, which have great coverage and increase interaction with audiences. It is also important to listen to people: a local farmer or radio producer knows everything about what makes people listen to or ignore communication messages and can teach you a lot about the local realities you're dealing with.

The most important thing when working in this field is to remain passionate. Remembering your initial reasons for wanting to work in this area and what you love about it will help you deal with the difficult times: securing a first job, dealing with frustrations once you have a job such as waiting for donor and management approvals, budget constraints and so on.

Rieke Weel (The Netherlands) India, Burundi,
Rwanda, DRC, Guatemala, Morocco

COUNTRY DIRECTOR

The Country Director (CD) is the most visible member of the organisation in-country and is often held liable for successes and/or mistakes in programme implementation. They oversee the programme and act as the communication link between HQ and the field office and usually report to a regional manager who will be in charge of a number of country offices in a large organisation. The CD is a strong multi-tasker, simultaneously dealing with internal management, programme implementation, business development and external relations on a daily basis. As a CD you should be ready to deal with a whole range of situations; and to address them successfully you might be ready to apply a combination of all your technical skills with a big dose of leadership, diplomacy and patience.

The work: Being a CD is a very diverse job: on any given day you can find yourself sharing breakfast in a remote village with one of the most vulnerable families and in the evening having dinner with the president of your host country. You will be responsible for managing and coordinating staff, developing and maintaining relationships with key stakeholders, strategic financial management, administration and logistical support, in-country strategic planning, proposal writing and budgeting, compliance with local regulations, overall security management and fundraising.

The employers: Any organisation that has a HQ-field office structure will have Country Directors (or a similar post) heading up their offices in their countries of operation. This includes NGOs, contractors and consultancy companies. Some employers may focus on a specific expertise such as human rights, WASH, food security and humanitarian assistance, whereas others have embraced a range of core themes in their approach for development.

Breaking in: this is a senior management role and being able to demonstrate competence in people management, diplomacy, financial oversight, report writing and strategic planning is essential, as will be direct experience in the type of programme the organisation is working towards. Besides a background in management and programme experience – ideally coupled with diverse technical knowledge – a CD must be an excellent communicator, working closely with colleagues at HQ as well as your employees, partners, donors, government officials and programme beneficiaries in the host country.

Job titles: Country Manager; Head of Office; Country Representative; Country Director

Find out more:

Dale Carnegie (2012) *How To Win Friends and Influence People*, London: Vermilion.
John C. Maxwell (2007) *The 21 Irrefutable Laws of Leadership*, Nashville, TN: Thomas Nelson.
Mark Alexander Williams (2001) *The Ten Lenses: Your Guide to Living and Working in a Multicultural World*, Herndon, VA: Capital Books.

See also: Programme Manager

I studied Civil Engineering in Honduras. After a few years of field experience as an engineer, I went back to school to earn a Master's Degree in Urban Planning in Switzerland. In October 1998 Hurricane Mitch hit Central America and Honduras became one of the most affected countries; social and economic infrastructure was severely damaged. I decided to take a sabbatical to volunteer my time in the reconstruction efforts. During this period I crossed paths with many International NGOs that had come to Honduras to provide humanitarian assistance and support reconstruction. A year later, without really looking for it, I joined the INGO I have worked for the past 13 years and have been constantly challenged since then.

I was initially hired as a housing manager and put in charge of an ambitious but successful housing initiative for families that had lost everything in the hurricane. The end of this programme marked the beginning of my international career. I first moved to Colombia to supervise an IDP programme. I was then named Country Director to Peru and sent there to set up a housing microfinance institution (MFI) in the Andean city of Cajamarca. I was a Country Director with no office, no employees and without knowing anyone in Peru. I arrived there with two briefcases and accounting software under my arm. In a year's time the MFI was operational, providing loans for low-income families; this operation was later turned over to locals who today continue to run the MFI successfully. From there I led the expansion of our operations in Bolivia in the field of alternative development, a term that I had never heard before.

After Peru I followed natural disasters: hurricanes in the Caribbean, earthquakes in El Salvador, Peru and Haiti, and the tsunami in Indonesia and Sri-Lanka, as well as post-conflict environments such as Afghanistan and Georgia, where I provided technical assistance programmes at all stages. I managed projects and programmes in a wide variety of fields including urban and municipal planning, alternative development, small and medium enterprise development, agriculture value chain, child protection, Malaria and HIV and AIDS programming.

I am now in Africa working as Chief of Party to a programme on orphans and vulnerable children, targeting over 360,000 beneficiaries. Our daily activities range from education and health to nutrition, food security, psychosocial support and economic strengthening to children, their parents and guardians. In addition I am also the CD, responsible for a team of 60+ people and eight implementing partners covering two-thirds of the country. In my work no two days are the same; each day brings its own challenges and fulfilments.

Being a CD is a big responsibility. My advice to anyone wanting to direct their career toward a higher level managerial position is 'go for it', prepare yourself, an MBA or a leadership qualification is a good thing to have. Speaking languages is also a plus in this career, along with your quota of willingness to explore and work in multicultural environments. As a CD you will never get bored.

Milton Funes (Honduras) Colombia, Peru,
Afghanistan, Georgia, Rwanda

CORPORATE SOCIAL RESPONSIBILITY (CSR)

'Corporate social responsibility is the commitment by business to contribute to economic development while improving the quality of life of the workforce and their families as well as of the community and society at large.'[1] CSR is an engine for social progress and helps companies live up to their responsibilities, using their power in business, acting in a socially responsible manner. For many companies it is becoming more than just an ethical duty, but is something that actually has a bottom line pay-off. Development NGOs are increasingly seeing that collaboration with the private sector is crucial for success. For instance, Visa partnered with Fighting Poverty with Financial Inclusion (FINCA) and USAID to expand access to microfinance and basic banking services to poor communities in Central America.

The work: You may be working within a company, helping to develop CSR policies and strategies. As a change agent you may be responsible for implementing these policies and changing corporate behaviour. More on the business side, you can open new markets by creating new sustainable practices or building sustainable business cases. Building partnerships between companies and NGOs, developing a joint agenda on social or environmental issues might also be a focus. Increasingly, multilateral organisations such as the UN and its subsidiaries are seeing corporate partnerships as a crucial instrument to meet development goals or create social change. CSR officials within multilateral bodies are actively building public private partnerships around common agendas.

The employers: There are three main types of employers in this field. Changing business from the inside, as a social entrepreneur, you might be working in a company's CSR department. Increasingly, as NGOs see companies as part of the solution, there are more roles within NGOs building corporate partnerships.

NGOs can engage with the private sector to help bring expertise in a particular field. Third, you can work with a consultancy company or think tank, helping companies move in the right direction.

Breaking in: Understanding business, but also a passion for social change, development and humanitarian issues is crucial.

Job titles: CSR Manager; Relationships Manager

Find out more:

Olufemi Amao (2013) *Corporate Social Responsibility, Human Rights and the Law: Multinational Corporations in Developing Countries* (Routledge Research in Corporate Law), Abingdon: Routledge.

John Elkington (1999) *Cannibals with Forks: Triple Bottom Line of 21st Century Business*. Oxford: Capstone Publishing.

Marc J. Epstein (2008) *Making Sustainability Work: Best Practices in Managing and Measuring Corporate Social, Environmental and Economic Impacts*, Sheffield: Greenleaf.

David Katamba, Christoph Zipfel, David Haag and Charles Tushabomwe-Kazooba (2012) *Principles of Corporate Social Responsibility (CSR): A Guide for Students and Practicing Managers in Developing and Emerging Countries*, Houston, TX: Strategic Book Publishing and Rights.

Wayne Visser and Jeffrey Hollender (2011) *The Age of Responsibility: CSR 2.0 and the New DNA of Business*, Chichester: John Wiley & Sons.

When I was young I had an interest in world politics and wanted to do something about hunger in Africa. After my studies in Political Sciences, I worked for a Dutch think tank. This was just around the time when CSR was starting to be talked about, and Greenpeace activists occupied Shell's Brent Spar. It was an interesting time with a change of mindset: companies were no longer the problem but part of the solution. This was a big insight, something that I found interesting and that I believed in. I realised that to work effectively in this field, I would have to gain inside knowledge of business. I joined a consultant firm that worked with many of the large companies, guiding them along the first steps of CSR. I got to see many companies from the inside: Heineken, TNT, KLM. Initially I saw that companies wanted external support with CSR, but soon they take it close to themselves, and manage it internally with a CSR team, especially when the management firmly backs it.

There are three main steps in the evolution of CSR. The *first* is People–Planet–Profit, moving away from business just as an asset of figures and incorporating social thinking. Integrating these soft areas into people's key performance indicators (KPIs) is challenging. The *second* step is understanding how the resources that a company has can be put to good use. For example TNT, a global express company, had a lot to offer and we helped them to create a partnership with the World Food Programme. During the tsunami they ran the logistics in Aceh airport and helped WFP with their warehousing and improving professional standards. This also helped the TNT staff to be part of a company that they were proud of. The *third* step is to help companies do something extra, with the society drawing up a common agenda to work towards. We see that companies are willing to do more for society, but it doesn't help when each NGO pushes its individual wish list. Through stakeholder dialogues a common agenda can be developed that also allows mutual learning. The first example of such a common agenda was the Access to Medicine Index, which successfully encouraged the pharmaceutical sector to improve the availability of medicine in developing countries.

After ten years in consultancy I decided I wanted to work inside an organisation. I joined the Access to Seeds Index where I now work. Indexes are a good way to develop a common agenda by which we measure companies and hold them responsible. Most farmers in developing countries do not benefit from decades of crop cultivation that have taken place around the world. They mainly rely on informal seed systems and farmer saved seeds. The Index encourages the world's leading seed companies to bridge this gap by shining a light on good practices and motivating companies to do more. As part of my role I do field trips to several countries to map the seed challenge.

I would advise anyone wanting to work in CSR to get business experience as you really need to understand businesses to be able to work with them in social responsibility. For many business people, numbers count and what gets measured, gets done. Speaking the same language is vital in this field.

Ido Verhagen (The Netherlands)

Note

1   www.wbcsd.org/DocRoot/hbdf19Txhmk3kDxBQDWW/CSRmeeting.pdf (accessed 30 January 2015).

DISABILITY

Disabled people are some of the most vulnerable people in developing countries, often stigmatised by the community. According to the World Health Organization (WHO), over a billion people in the world, or 15 per cent of the world's population, live with one or more disabilities.[1] This number is higher in lower income countries although most commonly the rate is underestimated due to a lack of accurate data. Whether as a result of trauma, or from an impairment from birth, there are many different conditions that can lead to disability. Furthermore, poverty and disability are inextricably linked. The opportunity cost of disability, limited access to health care, education, water and sanitation, and employment, can all lead to economic vulnerability and social exclusion. If you want to identify some of the poorest and most vulnerable members of any community you need look no further than those with disabilities.

The work: There may be some hands-on and direct work, but most will be programme management, advisory work, training and capacity building or mainstreaming disability into wider programmes. Much of the work involves helping organisations in the country have the resources to service the needs of those with disabilities better. The 'twin track approach to disability' is common: this means specific programmes for those with disabilities, but also including the concerns of people with disabilities as part of all mainstream projects.

The employers: There are essentially two different types of organisations doing work for those with disabilities in development. First, those that run specific programs to address the needs of those with disabilities: organisations such as Handicap International and CBM. But increasingly more and more agencies are moving towards mainstreaming disability issues into their existing programs, with multi-lateral and bilateral organisations such as UNICEF including this in their remit.

Breaking in: Any experience working face to face with people with disabilities is a bonus. A technical skill such as physiotherapy, occupational therapy or speech therapy is also a huge asset. These skills will help you better understand the service requirements that come with disability. However, from an inclusion perspective, more and more people without disability backgrounds are entering the field, often those that have experience with mainstreaming other vulnerable populations.

Job titles: Disability Advisor; Disability Co-ordinator; Inclusion Specialist; Rehabilitation Advisor

Find out more:

CBM's Inclusion Made Easy: www.cbm.org/Inclusion-Made-Easy-329091.php
End the Cycle: www.endthecycle.org.au
Malcolm Maclachlan and Leslie Swartz (2009) *Disability & International Development: Towards Inclusive Global Health*, New York: Springer.

After graduating and working as a physiotherapist in Australia, I realised that I no longer wanted to be a hands-on practitioner. Not knowing exactly what I did want to do, I spent two years living abroad, with part of that time spent volunteering in an orphanage and adult shelter for people with disabilities. I guess over that time I became more aware of poverty and interested in the causes of it. I began to see disability in a different light. Upon returning to Australia I began working with an NGO that made custom-designed assistive devices for people with disabilities. Concurrently, I completed an MA in Development Studies at the University of New South Wales. I knew that I wanted to work in development, but I didn't know what my focus would be. When I was offered a job at Handicap International, working in China as a Rehabilitation Advisor, I knew disability was the area I wanted to be in.

Most of my recent work has been in supporting Cambodian grassroots NGOs in helping those with disabilities to participate better in society and help these organisations to be more effective. This involves balancing out the needs of the community and capabilities of the organisation, with the technical knowledge and experience that I have. It is 90 per cent listening and the rest speaking. The work these grassroots organisations do is so important because the need is so great, and there are very few service providers. Staff often spend hours on motorbikes enabling families in remote communities to access services that have the fundamental goal of improving participation and independence for those with disabilities. This may mean integrating children into schools, helping them to walk independently, or ensuring that elections are accessible for all people. It can be challenging working with organisations of this size. For one, they do not have anywhere near the budget that large multilaterals do. For instance, some UN agencies would spend as much on one conference as a small organisation would spend on an entire programme annually. However, these organisations generally do the best they can with the few resources they have.

More broadly, working in disability is incredibly challenging because despite much rhetoric, it is still not an issue that has centre stage in development. There was no specific mention of disability in the Millennium Development Goals or in the Post-2015 Development Agenda. This means you are constantly fighting for it to be recognised, especially among donors. This is not to say that things are not changing, or that working in this field is unenjoyable; quite the opposite. Simple things can have a big impact on people's lives. For example, a wheelchair and physiotherapy services provided to a person who has never left their house for many years, can be transformational.

In an attempt to have disability more widely recognised in the development sphere, I try and write about the importance of this topic on WhyDev (http://whydev.org), a site I co-founded to help promote discussion and collaboration on aid, development and global issues.

For those who want to work in disability, my advice would be to spend some time with those with disabilities first. Get to understand what the real needs are, and be prepared to listen. Have patience, because nothing will come quickly. Be prepared to fight and hustle for the cause. You will have to spend much of your time justifying why this isn't an issue that deserves to be ignored.

Weh Yeoh (Australia) Vietnam, India, China and Cambodia

Note

1   www.who.int/mediacentre/factsheets/fs352/en/ (accessed 30 January 2015).

DISASTER RISK REDUCTION (DRR)

The number of people affected by disasters almost doubled in the last decade.[1] Global climate change, environmental degradation and the fast pace of urbanisation are, among other factors, placing more people at risk of disasters. The high human impact of the 2012 floods in Pakistan or the earthquake in Iran are evidence of poor land management, infrastructure or planning. Humanitarian aid needs to be coupled with risk reduction initiatives to lessen the impact of natural hazard events. Integrating disaster risk reduction (DRR) within the development agenda is an imperative; development partners are increasingly focusing on investing more in DRR as it becomes apparent that addressing – and financing – it as an afterthought is not good enough. Research indicates that every dollar spent on minimising risk saves seven in economic losses due to disasters (World Bank, 2004).

The work: The field of DRR bridges the humanitarian and development communities. You might be working in high-risk locations helping local communities and authorities to build their capacity to respond to disasters through the creation and training of disaster preparedness teams. You could also be addressing national legislation including development of policy and working instruments such as guidelines and manuals that enable key sectors to integrate risk analysis and risk reduction measures into their plans. Another role is within public information, providing early warning messages or education on disaster risk and how to reduce it at the household or community level. Addressing underlying issues that directly impact disaster risk, such as water resource management or urbanisation is another focus.

The employers: The UNDP supports more than 60 countries in creating national disaster risk reduction strategies and announced in 2012 that it is doubling DRR over the next five years. The United Nations International Strategy for Disaster

Reduction (UNISDR) is the UN body responsible for driving the risk reduction agenda and for building coordinated efforts among multiple stakeholders. Many other development organisations work in DRR, including Oxfam, CARE, World Vision, CRS, Save the Children Fund (SCF), Mercy Corps, Christian Aid, to name but a few.

Breaking in: Knowledge of the humanitarian context helps you to understand the consequences of disasters, making prevention and risk reduction a logical step. The key is to understand what makes development sustainable and resilient to shocks, such as the occurrence of natural hazards, and then to apply those principles in a context of high disaster risk, to enable people and communities to build resilience. There are Master's degrees in Disaster Risk Management, for those who prefer to study a subject before attempting to tackle it.

Job titles: Disaster Risk Reduction Advisor/Coordinator; Resilience Advisor/ Coordinator; Adaptation and Risk Reduction Manager

Find out more:

Piers Blaikie, Terry Cannon, Ian Davis and Ben Wisner (2004) *At Risk: Natural Hazards, People's Vulnerability and Disasters*, New York: Routledge.
Marilise Turnbull, Charlotte L. Sterrett with Amy Hilleboe (2013) *Toward Resilience: A Guide to Disaster Risk Reduction and Climate Change Adaptation*, Rugby: Practical Action Publishing.
www.preventionweb.net

I started my career working as a project officer for a small child-centred development organisation, during which time I learned about project cycle management, sustainable development principles, and how to work with multiple stakeholders on common aims. Then I moved into the humanitarian sector, working in Colombia with IDPs displaced by conflict, and with people affected by a major earthquake in another region of the country. For the following six years I worked as a Humanitarian Coordinator for Oxfam, undertaking numerous assessments, writing proposals, building staff capacity in preparedness for response, and evaluating programmes. It was during this time that I became interested in disaster preparedness, and then disaster risk reduction. I undertook many assessments of humanitarian crises, and studied the evaluations of other humanitarian programmes, and started to feel that it did not make sense to wait until a disaster occurred to address the causes – which could be seen clearly from well before. I wanted my contribution to be about bridging a gap between development and humanitarian practitioners, both of whom see DRR as in the other person's camp.

I was asked to lead Oxfam's work on disaster preparedness, which rapidly transitioned to disaster risk reduction – an emerging field at that time. After several years of internal advocacy we managed to create an Adaptation and Risk Reduction team in Oxfam, which I managed. In 2008 I became an independent consultant, and currently I support organisations with policy writing, programme development, capacity building and research into aspects of disaster risk reduction and climate change adaptation.

Currently I undertake a wide variety of tasks for international organisations. Today, for example, I interviewed people from six countries to find out what sort of advocacy for DRR could generate the best results in their area. Based on this and other sources of information, I will put together a global proposal for Oxfam.

There are many useful combinations of academic studies to work in this field. I have an MA in Modern Languages and an MSc in Development Management. Both have helped me gain communications skills, investigative and analytical capacity, and understanding of development processes (successes and failures). But to work in this field you could be a food security specialist, a geologist, a social promoter, a public health technician, or have another related area of expertise.

I like the variety of the work related to DRR and the conceptual versus practical challenges. But I don't like having to seek difficult-to-find funding for risk reduction when it seems so obviously an area in which all stakeholders should invest their time and resources.

If you want to work in this field, make sure you understand climate change and adaptation. Keep broad to start with, so that you do not miss any interesting subjects. Get hands-on field experience by volunteering or finding a paid position. To develop my knowledge I read a good deal about others' experiences of DRR across all the sectors. There are of course many related courses and seminars, many of which are listed on Preventionweb (website).

Marilise Turnbull (UK) Colombia, Peru, Haiti, Kenya, Cambodia

---

Note

1   The number of people affected by extreme natural disasters, increased by almost 70 per cent, from 174 million a year between 1985 to 1994, to 254 million people a year between 1995 to 2004. Floods and wind-storms have increased from 60 events in 1980 to 240 in 2006, with flooding itself up sixfold. But the number of geothermal events, such as earthquakes and volcanic eruptions, has barely changed (Centre for Research on the Epidemiology of Disasters (CRED)).

ECONOMICS

Development economists look at the economic aspects of the development process in low- and middle-income countries and how both micro- and macro-level policies can promote poverty reduction. It has become one of the liveliest areas of research of all the social sciences, drawing on economic theory, econometric methods, sociology, anthropology, political science and demography. Given that development often focuses on reducing poverty, economists play a central role in poverty reduction strategies and methods for promoting economic growth and structural change, as well as how to improve public access to health, education and employment. Emerging fields such as behavioural economics are also gaining importance.

The work: At the macroeconomic level, you may be influencing government policies to improve the way markets work, encouraging rapid and sustainable economic development or supporting one of the development banks/donors in the design and implementation of their programmes and projects to support poverty reduction. You could also be exploring issues such as poverty, inequality and corruption, or understanding taxation and how governments can generate their own revenue. From a microeconomic perspective, you might be looking at what poor people spend their money on, microfinance and boosting poor people's incomes.

The employers: Some of the obvious ones are the international finance institutions (World Bank, IMF, African Development Bank, Asian Development Bank). But economists are in great demand in many bilateral donors (e.g. DFID, USAID), other multilateral institutions (e.g. the UN), development consultancy firms and think tanks. Many economists will specialise in a particular area, for example the WHO will employ health economists to improve health care financing.

Breaking in: A Master's degree in Economics (a development focus is not a prerequisite but can be an advantage). You don't always have to have an undergraduate degree in economics to study it at Master's level, but you will need some mathematical background. A doctoral degree can be important for some employers. Some bi- and multilateral organisations have recruitment rounds specifically aimed at entry-level positions for economists (e.g. the World Bank Young Professionals Program). The UK's ODI fellowship scheme recruits postgraduate economists to work in the public sector of developing countries. They welcome applications from all nationalities and in 2012 they received around 600 applications for 54 posts.

Job titles: Development Economist; Economic Advisor; Health Economist; Energy Economist; Natural Resources Economist

Find out more:

Abhijit Banerjee and Esther Duflo (2011) *Poor Economics: Barefoot Hedge-fund Managers, DIY Doctors and the Surprising Truth about Life on less than $1 a Day*, London: Penguin.
Jagdish Bhagwati and Arvind Panagariya (2013) *Why Growth Matters: How Economic Growth in India Reduced Poverty and the Lessons for Other Developing Countries*, New York: Public Affairs Publishing.
E.F. Schumacher (1991) *Small is Beautiful: A Study of Economics as if People Mattered*. London: Abacus.

See also: Governance; Microfinance; Livelihoods

After completing my undergraduate degree in Economics and Management at Oxford University and working for an economic consultancy in London, I chose to study for a Master's in Economics at the London School of Economics (LSE). It was during this time that I decided to pursue a career in development. I had always been interested in travel and I wanted to combine that with a job that used my economics skills to genuinely improve people's lives.

While doing my Master's I applied for the ODI Fellowship Scheme and was selected to work in the Ministry of Finance in Guyana. The fellowship is a two-year placement and is one of the best ways to get started working in development economics. You work as a civil servant in a developing country and are line managed directly by staff in your Ministry. This provides a fantastic opportunity to really understand how governments in developing countries work and to support and influence their policies for economic growth and poverty reduction.

As an ODI Fellow, one of the things I worked on was debt cancellation. It was around the time of the G8 Summit at Gleneagles and the Government of Guyana was supporting the campaign to cancel debts owed by the world's poorest countries to the IMF and World Bank. We also worked to extend the debt cancellation to cover the Inter-American Development Bank, resulting in the cancellation of hundreds of millions of dollars of Guyana's debt. It was an exciting time and great to see the impact that the cancellation had on Guyana's ability to fight poverty.

After Guyana I returned to the UK and decided to work for a development consultancy. Development consultancies often implement projects on behalf of bilateral donors and multilateral institutions, so the work provides the opportunity to continue engaging directly with governments in developing countries, as I did as an ODI Fellow. I work for Oxford Policy Management (OPM) and now lead the Public Financial Management team.

We typically work on several projects at any one time – providing technical advice and training to support reform programmes and build capacity in government institutions. For example, we are currently working with the government of Uganda to develop an integrated macroeconomic model of the Ugandan economy. We are training staff in the Ministry of Finance to use the model to forecast the economy, determine economic policy and set the government budget.

Once you have a Master's, the ODI Fellowship Scheme is one of the best for those wanting a career in the field. It places young postgraduate economists in the public sectors of developing countries for two years (www.odi.org.uk/fellowship-scheme).

Ed Humphrey (UK) Guyana, Uganda, Pakistan, Mongolia, Zambia

EDUCATION

Education lies at the heart of development and poverty reduction. It is a means to achieving sustainable improvement in livelihoods and well-being, and directly correlates with health outcomes, democratic processes, population growth, improving incomes and more. With the second Millennium Development Goal (MDG) focusing on primary education, significant investments have been made to improve access to education, including the abolishment of school fees, investing in teaching infrastructure and promoting education for girls. However, it is unlikely that this MDG will have been met by the target date 2015, especially in Africa and Asia where school drop-out rates remain high. But education is not just a development issue and has recently gained in importance within emergency responses. Access to *quality* education remains a challenge across the board.

The work: Education in development and humanitarian contexts is multifaceted and the scope of the work is vast. Areas of focus include children's equitable access to education, improving the quality and relevance of education, best practices in literacy and numeracy instruction, teacher development and training, school health and nutrition, school management, education economics, technology for education, improving access to basic education for disadvantaged children, as well as capacity building at government level, implementing education policy or improving access to education during emergencies. Many donors' emphasis is on literacy, early grade teaching and girls' education, but early childhood development is also gaining importance.

The employers: As education is a core component of development, the number of organisations employing education experts is significant. Multilateral and bilateral organisations often have *education advisors* to inform their programming. Private consultancy companies, countless NGOs and humanitarian organisations also have education programmes.

Breaking in: A teaching background can be an advantage – especially if you want to go into teacher training – but is not essential: economists, peace builders and artists may also be found in the field. A Master's in Development Education will help launch your career. Developing a sound background in at least two areas, such as teacher education, educational measurement, curriculum development, textbook writing, IT, school mapping or education economics will be an advantage. Knowledge of impact evaluation methods is also important.

Job titles: Education Specialist; Education Advisor; Early Childhood Care and Development Specialist; Education in Emergencies Specialist

Find out more:

Education for All (EFA): www.unesco.org/new/en/education/themes/leading-the-international-agenda/education-for-all/
Join the LinkedIn group Education in Developing Countries
Paulo Freire (1968/2000) *Pedagogy of the Oppressed*, London: Bloomsbury Academic.
The Inter-Agency Network for Education in Emergencies (INEE): www.ineesite.org
The International Institute for Educational Planning (IIEP): www.iiep.unesco.org
Zuki Karpinska (2012) *Education, Aid and Aid Agencies (Education as a Humanitarian Response)*, New York: Continuum.

See also: Orphans and vulnerable children (OVC); Project management

---

For my Master's in Social Anthropology, I conducted five months of fieldwork in a village in Bhutan. This experience strengthened my long-standing interest in development and the education sector in particular. Getting my first job was difficult, however. I am a trained social worker and had about five years of work experience in child protection and education. I did not have any development experience though. It was a vicious circle that I broke out of by doing an internship at UNESCO in Nepal, working on education and communication and information. This opened many doors, especially since I decided to stay on in Asia. Because of my experience with the context and through my network, I was able to take on a number of assignments for different NGOs and UN agencies, mostly in the education sector.

After almost two years, I moved home when I was offered a PhD position to study how village-level leadership in Nepal is influenced by the democratisation process. I did not want to lose my link with the development sector though, which is why I applied to NORCAP, the emergency standby roster of the Norwegian Refugee Council. In 2010, on a couple of weeks' notice, I took leave from the university to be seconded to the UN's Children's Fund (UNICEF) in Haiti. For six months, I worked with the coordination of education as an emergency response. This was my first time working in an emergency context. It was an extremely interesting and rewarding experience.

In April 2012, I took on another three-month NORCAP assignment, this time in Niger, where I again worked for UNICEF. My role was to coordinate all the education in emergency initiatives in the country, in collaboration with the government. The education response focused on two major issues: the 20,000+ Malian refugee children in Niger who needed education, and the food crisis affecting the entire Sahel region, causing more than 50,000 children in Niger to temporarily quit school. We focused on training the (inter)national aid organisations in Niger on the particularities of education provision in emergency situations. We developed guidelines to minimise the consequences of schools being used as shelters, which is common when disaster strikes. We also worked to set up temporary schools in the camps, run by the 30+ refugees who were teachers, school directors and pedagogues.

After this assignment, I joined a Global Education Cluster Rapid Response Team responsible for the coordination of education in emergencies, a secondment to UNICEF through NORCAP. In this position, I am in the field 60 per cent of the time, and provide home-based support the rest of the year.

As an anthropologist, I particularly like the work in and with the field. Education defines children's future, and it is often rated as a priority by the communities themselves. I therefore find it an extremely fulfilling aim to contribute to.

Annelies Ollieuz (Norway) Bhutan, Haiti, Nepal, Niger, Vietnam

ENERGY, ENVIRONMENT AND CLIMATE CHANGE

Environmental specialists with a background in development are in demand these days, especially since the Kyoto Protocol came into effect in 2005 and climate change began becoming more visible on the world agenda. Promoting renewable energy sources in the developing world has become a key focus for many NGOs and governments who are now supporting climate change mitigation, adaptation or resilience projects. Financing for carbon offset projects/programmes is also an area that has grown quickly, and that has the potential to benefit the private sector and governments in the developing world by creating revenue from selling carbon credits. Access to energy – and renewable energy – is also of critical importance for sustainable development.

The work: You will be working toward *mitigating* (e.g. promoting renewable energy, energy efficiency or forestry projects) or *adapting* (e.g. designing improved agriculture programmes that are more resilient) to climate change. The focus might be on resource and energy efficient technologies, waste management or low emissions development strategies. These projects can be challenging to implement as they seek to modify current practices – such as replacing fossil fuel with renewable energy, reducing consumption of non-renewable biomass through the use of improved cookstoves or replacing kerosene lanterns with solar lanterns – and carbon revenues are only generated post-implementation and post-monitoring. You might also be undertaking financial modelling for funders and equity investors, engaging in biomass-to-energy projects or wind and hydro power projects and the supporting grid infrastructure.

The employers: Many NGOs, both local and international are focusing on these issues, as are many social enterprises such as Nuru Energy (http://nuruenergy.com). In addition, most, if not all, development agencies and bi- and multilateral agencies are financing climate-related activities. For example, the UNDP in collaboration with the European Union (EU) has developed the low emissions capacity building project which is part of the global partnership programme implemented in 25 countries. Others might include United Nations Framework Convention on Climate Change (UNFCCC); United Nations Environment Programme (UNEP) or the Department for Energy and Climate Change (DECC).

Breaking in: Some people come into the field with a background in environmental sciences or engineering, while others learn on the job and take some short courses related to this field or a Master's such as in Environmental Management. The environment sector is so broad that it needs people from all backgrounds – scientists, engineer, lawyers, economists, social scientists. It is important to learn about and understand the challenges, opportunities and complexities of the relevant environmental issue(s).

Job titles: Renewable Energy Specialist; Climate Change Policy Advisor; Environmental Economist; Hydrology Engineer; Soil Technician; Agriculture Advisor; Lawyer with a focus in environmental law

Further information:

Climate Development Knowledge Network, www.environmentjobs.com/
Rogers *et al.* (2007) *An Introduction to Sustainable Development*, London: Routledge.

See also: Disaster risk reduction; Livelihoods; Food security

---

Until recently, many people working in climate change entered the field later on in their careers but I became involved in climate change early on in my undergraduate university days. At this time, it was before climate change had become a 'trendy' topic and many questioned how I would ever find a job in this field. I completed my Bachelor's degree in Environmental Science and Management at the University of Rhode Island. During this degree, I became interested in climate change while working on a paper on the Kyoto Protocol; this was in 2003, two years prior to the Protocol entering into force.

After this, my interest in climate change peaked, and I completed a number of climate change related internships; doing these internships was very useful for exposing me to a wide variety of aspects of climate change and for kick-starting my career. I did internships with an NGO in Australia working on climate change outreach; with the US Environmental Protection Agency; with a wind project developer; and with the Atmospheric Chemistry Department of my university. I then moved to Europe and undertook a European Master's programme in Environmental Management as Europeans were (and still are, at an upper governmental level) much more progressive than Americans on the subject of climate change. Following the completion of my Master's degree, I joined a carbon market company called EcoSecurities, focusing on emissions reductions projects primarily in Brazil, Southeast Asia (mainly Indonesia, Malaysia and Thailand) and China.

I then decided that I needed to reduce my emissions from travelling and that it was time to be based in one of the developing countries that I travelled to so often. I secured a role as Carbon Market Technical Advisor to the Government of Rwanda through the UNDP for one year. My experience working with the Government of Rwanda was a fantastic one. It was a pleasure to work with a very progressive and motivated government and group of people.

I decided to stay in Rwanda after finishing my UNDP contract and I worked as a climate change consultant. I worked on a DFID project using results-based financing to promote private sector involvement in renewable energy, on a conservation project and assisting with a number of carbon credit projects including water purification and with cookstoves.

Finally, after almost three years, I decided to join a climate change consultancy firm in Zurich. I work with a variety of clients including government aid agencies (e.g. GIZ) and multilateral agencies (e.g. UNDP) and I have maintained a strong focus on projects in East Africa. I also started a part-time PhD in Political Science and Economics, with a focus on climate change. For part of my PhD research, I am looking at climate finance projects in Rwanda that will provide clean water and improved cooking conditions to rural Rwandans.

Climate change is a very exciting area, especially as the scope of it is so wide. It is important to figure out in which branch (e.g. mitigation, adaptation), component (e.g. water, forestry, energy), sector (e.g. private sector, government, non-governmental) and level (e.g. local, national, international) you want to work. Completing internships is an excellent way to help you figure this out.

Courtney Blodgett (USA), China, Rwanda,
Brazil, Indonesia, Thailand

ENGINEERING

Engineering is an important component of humanitarian response and sustainable community development. One of the challenges in many developing countries is poor infrastructure, access to energy (with a particular focus on renewable energy sources) and developing local solutions to problems, all areas where engineers can play important roles. During reconstruction efforts, the focus will be on rebuilding and improving upon previous structures. Areas in demand are renewable energy and environment (especially energy efficiency and clean energy), civil engineering, construction and computers/ICT. Engineering also serves as a good background to other types of work, as there are many transferable skills, especially for numerical, analytical and survey type roles.

The work: Similar to your work back at home but in a different context, with a different level of technology and in a different language. Some additional responsibilities as well: it might involve project management, elements of teaching and training local staff, transfer of knowledge and skills and so on. You might be developing a windmill using local materials as an energy source for a rural, off-grid community, building a school in a war-torn area or empowering people to use water effectively. Several university teaching jobs (to train up local engineers) are also available.

The employers: Engineers without Borders (EWB) is a good place to start. But many larger NGOs or international agencies involved in reconstruction or development work may employ engineers at different points in time, according to projects, especially if they already have experience of development projects. International consulting firms are often contracted to deliver this type of work as well.

Breaking in: Specialise in your particular field of engineering, get at least three to five years of private sector experience. Before looking for a paid position get some experience in working in a 'development' context. Volunteer work is a good way to do this. EWB work with disadvantaged communities and have several member groups and chapters in several different countries; find your closest hub on EWB International (www.ewb-international.org). EWB often recruit volunteers for placements around the world and, typically, living allowance, flights, accommodation and support are covered.

Job titles: Engineer; Team Leader; Project Manager; Construction Manager; Field Officer

Find out more:

William Kamkwamba and Bryan Mealer (2010) *The Boy Who Harnessed the Wind*, London: HarperTrue.

Juan Lucena, Jen Schneider and Jon A. Leydens (2010) Engineering and Sustainable Community Development. Synthesis Lectures on Engineers, *Technology and Society*, Vol. 5, No. 1: 1–230.

See also: Shelter and housing; Urban planning

I had always been interested in working internationally, but after graduating I had debts to pay off, and so needed a steady, private sector job. I worked on various projects in Hong Kong as a structural engineer and later, with more experience, in project management. When I turned 40 I needed a change. The work I was doing wasn't very interesting or challenging to me anymore and I had financial stability, so I quit my job. In 2005 the Hong Kong Red Cross was looking for a construction expert to be part of a team rebuilding housing in Sri Lanka to replace those destroyed by the tsunami. I was lucky enough to get the position.

Later I worked with UNICEF in Banda Aceh, a position I applied for through ReliefWEB. I was recruited as the Construction Manager. It was the last year and a half of a five-year project building about 350 schools and 150 clinics. I managed a team of around 30 staff in three field offices, of which around 20 per cent were international. We had around 100 contracts to manage with local contractors who did the building work. The rest of the team dealt directly with the contractors on a day-to-day basis while my main tasks were ensuring systems were in place and in accordance with agreed processes and to the required standards, signing off payments, finding solutions to problems beyond the capacity of other team members and dealing with internal issues to get approval and so on. I occasionally went on site visits, but distances were so large, and the sites so remote, it would take a whole week to visit just a few.

The actual day-to-day work in Banda Aceh wasn't much more exciting than my previous work back in Hong Kong but I enjoyed it more because of the extra dimensions: first, working in a different culture, and with people from various countries;

second, the people would benefit from the project – instead of working on a project making money for a developer, I could see that the end result of our work would improve the lives and standard of living of the beneficiaries, e.g. children going to their new and better quality schools. It was more satisfying.

After Banda Aceh I got an assignment as a consultant in Africa, also with UNICEF. My advice to any engineer interested in this field is that after graduation, first get private sector experience, and boost this with international volunteering placements. Ideally, gain some financial stability and then start looking to work in this area. Technical competencies aside, I think the most important quality is flexibility. It would be an advantage if you can demonstrate that you have successfully adapted to different working environments in the past. I had the unfortunate experience of working with someone who was always homesick and did not adapt well on the mission. Not only did it impact on the performance of this individual, but also adversely affected other colleagues who were already working in a stressful and difficult environment. Situations like this can bring the whole team down.

Silla Chow (Hong Kong), Sumatra, Sri Lanka, DR Congo, Jordan

FARMING AND AGRIBUSINESS

Agriculture is a way of life for many of the world's poorest, three-quarters of whom rely either directly or indirectly on farming for their livelihoods. Agricultural development can make a difference in the lives of more than a billion people – mainly family farms where women play an important role – and can contribute to economic growth, poverty reduction, food security and reduced urban migration. Increasing agricultural yields and facilitating access to markets are important focus areas. Integrated soil fertility management, access of smallholder farmers to quality seeds and credit are key to raising yields per land unit. Collective action of small farmers is important for professionalisation, procurement and marketing. For cash crops and food crops for national and regional markets, linking of farmers to (often distant and non-transparent) markets is a priority for feeding 9 billion people in 2050. Since the Green Revolution, total agricultural output has risen, but in many countries production has not kept up with population growth, creating an issue of food security and land disputes. Climate change will have a far-reaching impact on agriculture and mainly hit less well-endowed farmers.

The work: On the technical side, fields of specialisation are crop variety development, soil fertility management, pest and disease control, efficient agricultural practices, water management and irrigation, mechanisation, harvest and storage loss reduction and others. On the financial side, the innovation of rural finance and agricultural insurance are important fields of work. On the economic side, you might be focusing on product development and agro-processing, supply chain management, branding and other subjects. On the policy side, the work might focus on trade barriers, agricultural subsidies, land rights, revenue generation from selling carbon credits and other strategic subjects. In the field of research and education, action research and training and coaching of farmers and other rural entrepreneurs remain important fields of work. The organisation and professionalisation of (smallholder) farmers is a cross-cutting subject.

The employers: The list of potential employers is extensive, ranging from multilateral organisations (such as IFAD, World Bank) to bilateral aid agencies (such as GIZ or the Belgian Technical Cooperation (BTC), consultancy firms, knowledge centres and universities, and to a wide range of international and national NGOs. In the agricultural domain, it is also increasingly important to consider large private companies as potential employers as supply chains are globalised. Experts in the agricultural domain can also set up their own business, for instance in the form of joint ventures with local entrepreneurs.

Breaking in: Formal studies in agronomy, (agri)business management, rural development and economics, with a developing country focus will be a good start.

Potential job titles: Agriculture Advisor; Agriculture Innovation Manager; Agriculture Extension Specialist; Agricultural Analyst

Find out more:

Chris Penrose-Buckley (2007) *Producer Organisations: A Guide to Developing Collective Rural Enterprises*, London: Oxfam GB.
Chris Reij and Ann Waters-Bayer (eds) (2013) *Farmer Innovation in Africa: A Source of Inspiration for Agricultural Development*, London: Routledge.

See also: Value chains; Rural development; Livelihoods; Food security; Microfinance

---

I studied Tropical Crops Science at Wageningen University in the Netherlands. My international career started as a VSO Irrigation Agronomist in northern Nigeria. Stationed at the Kano State Rural and Agricultural Development Authority my task was to try new crop varieties in farmers' fields. After two years, I moved to a wetlands conservation project in the same region where I coordinated micro projects to improve the use of wetlands resources.

I then returned home to find that Netherlands-based jobs in the field of development were scarce. Not yet ready to retrain, I applied for a position with Winrock International to coordinate a rice technology transfer project in Ivory Coast. Based at the West Africa Rice Development Association (now the Africa Rice Center) I learned a lot about rice in Africa but also that research is not really my thing. After another long-term position in Mali to manage Winrock's agriculture programme I moved back to Nigeria to join a USAID funded agribusiness project called MARKETS to head a team of agribusiness specialists. With a very strong emphasis on market-led value chain development, this project offered a tremendous opportunity to get hands-on experience managing partnerships with national and international agribusiness firms.

The key focus of the project was to assist these firms with the sourcing of raw materials through the development of outgrower programmes. My earlier experience as a VSO volunteer turned out to be very useful. The project I headed provided technical support in the areas of extension, credit mobilisation and processing. A typical day in this role would start by discussing key priorities for that week with the team leader. After checking emails I might meet with the representative of a large rice mill to review the memorandum of understanding. In the afternoon there could be a workshop on Nigerian agriculture where I would give a presentation for an audience of bankers and agri-entrepreneurs on demand-led production through value chain development. The interest in agriculture as a business is steadily growing in Nigeria, but there are still many risks and challenges.

In 2009 I moved back to Europe to join the European Cooperative for Rural Development (EUCORD), a Brussels-based non-profit organisation affiliated to Winrock. Here I continue to support farmers to access markets, through partnerships with companies such as Heineken and Diageo (Guinness). One of the things I enjoy most about my work is to hear farmers testify about the impact of the project on their lives. The ultimate reward is when activities are continued or even replicated well after the project has ended.

My background in agronomy gives me confidence in the field, but on a day-to-day basis I am a manager. Agribusiness development is multifaceted and a business, economics or development degree can equally carry you a long way. One has to be ready at all times to learn, never leaving behind good common sense.

Niels Hanssens (Netherlands) Burundi, DRC,
Ivory Coast, Mali, Nigeria

FINANCE AND ACCOUNTANCY

The finance manager performs a central role in any organisation and it is an area that will always be in demand. You could be overseeing the entire organisation's finances, managing day-to-day finance within a country- or regional-level operation, or working on a specific project's accounts. The dynamics of working in accountancy or finance for an NGO can be quite different than for a private company. Although suppliers will still require to be paid, and financial controls enforced, your income doesn't come from sales but from donations, grants or contracts and each donor will have their own reporting requirement. They will often also include specific limitations on which items they are prepared to fund, and on how suppliers must be selected. Instead of shareholders you are working for the beneficiaries and will be reporting to trustees.

The work: If you are working on a specific project you will be responsible for all accounting tasks associated with it. You might be applying accounting procedures to programme structures, reconciling accounts, processing and reviewing project financial records, preparing monthly invoices to funding agencies and preparing project financial reports and budgets. Building capacity in local partners' organisations is necessary, so you could also be training them in relevant simple accounting systems and packages such as QuickBooks, establishing the financial systems required by donors, or auditing their accounts. At a higher level you might be working with the management to discuss income flows and forecasts or planning the organisation's global budget. In smaller field offices you may find yourself managing logistics or related administration activities.

The employers: All organisations that handle any money are in need of accountants and finance directors, from the large multilateral organisation right down to the smallest community-based organisations. The nature of the work will vary depending on the organisation you work for.

Breaking in: A finance and accountancy background (a degree, a Chartered Institute of Management Accountants (CIMA) qualification or an Association of Chartered Certified Accountants (ACCA) qualification) is essential for a full-time finance position. Most people going through this route work for the private sector for a few years to gain experience in the area, and then move across to the not-for-profit field. Many larger organisations will require you to have knowledge of donor reporting requirements (such as USAID, DFID, EU grants) but flexibility and willingness to work in challenging environments is important.

Job titles: Finance Manager; Finance Controller; Project Accountant; Regional Accountant

Find out more: Management Accounting for Non-Governmental Organisations (MANGO) www.mango.org.uk advertises jobs, offers financial training and a handbook on financial management essentials for NGOs.

See also: Economics; Governance

I studied History through to Master's level, and then joined a small accounting practice in the UK. This gave me a thorough grounding and enabled me to gain professional qualifications recognised worldwide. I worked in Luxembourg, then Germany for commercial firms before spending several years with an oil company, as a finance manager in Nigeria, Russia and Oman. Although I enjoyed the challenge of adapting to new environments and cultures, I realised that I no longer wanted to work in the commercial sector. Through research I learned that many of the finance management skills were not only transferrable to the humanitarian sector, but very much in demand.

The main challenges in making this move were financial – I took a significant pay cut from an oil company to my initial role with a local community-based organisation in Kenya – and in starting over with learning about reporting, since the needs of an NGO are very different from a commercial organisation as their major donors (USAID, ECHO, DFID etc.) have very different priorities than shareholders. Plus of course, as the new individual in an organisation I had to start at the bottom again, despite my previous experience.

My first role was tough – I was the only expat in a remote area of Kenya and although people were friendly and helpful, it was sometimes lonely. I learned a lot in a few months and could see the impact of my organisation and my own efforts on the lives of people around me. I was also surrounded by beautiful nature – I've always loved wildlife – and even on difficult days I knew many would gladly swap places with me. After one year I moved to a larger organisation, spending six months in an acute humanitarian role in Gaza. Since then I have worked in Afghanistan, Haiti, Darfur, Niger and Bangladesh for organisations including Merlin and Mercy Corps.

I now work for Concern Worldwide and am based in Uganda. As a Regional Accountant I provide financial oversight for six countries in Central Africa (Burundi, DRC, Niger, Rwanda, Tanzania, Uganda). Across the region we have around 20 finance staff. This presents challenges such as language for example – if you cannot read the local language your control is diluted. More difficult is the fact that years of internal conflict in some countries have resulted in very low standards of technical training, so that the level of local staff varies.

The technical accounting skills needed in most NGOs is often fairly limited – in a field office at least. It is often more about being organised, structured, having systems and policies and effective controls. A typical month will include a support visit to one country (to work with the team on staff development, review of controls, systems support etc.), as well as a review of donor reports for all countries in the region. I also work on reviewing budgets and reports for both internal users (Country Directors, Regional Directors etc.) and for submission to donors.

One day I will return to the UK to be closer to family, but I will take with me a treasure trove of memories and friendships, and the satisfaction of having made a contribution to improving the lives of others.

Phil Crosby (UK) Uganda, Gaza, Afghanistan, Haiti, Bangladesh

FOOD SECURITY

Food security exists 'when all people at all times have access to sufficient, safe, nutritious food to maintain a healthy and active life'.[1] It hinges on *availability* (sufficient quantities of food available on a consistent basis), *access* (having sufficient resources to obtain appropriate foods for a nutritious diet) and *use* (appropriate knowledge of basic nutrition as well as adequate water and sanitation). Food security fell off the international agenda in the 1990s after the success of the Green Revolution, but as world population increases the challenge of food security has become very political and closely linked to economic development, environment and trade. Food security is vital to sustainable development and closely interlinked to issues such as property rights – which gives farmers the security that they will have their land long enough to realise the benefits of their investments.

The work: You might be working on research analysis or policy recommendations that can effectively enhance global food security. One example is the causes of food price volatility and the impact that this has on access. Another area might focus on cross-border trade and market opportunities. Implementing on the ground you might be working with smallholder farmers to build resilience to natural disasters and climate change, support the use of advances in technology to boost yields or support farmers in partnering with the private sector to get these crops to market and promote access. You could be helping a country build up its food reserves to ensure that enough is available on the market when a crisis hits, or working to develop secure property rights for farmers. In an emergency context the focus is likely to be on provision and meeting the affected population's basic needs.

The employers: From the World Food Programme (WFP) to the global food security initiative spearheaded by President Obama called Feed the Future, many donor programmes now have an emphasis on food and several NGOs are implementing projects in this area, including FAO, USAID, DFID, IFAD, UNIDO, World Bank, private sector and philanthropic organisations.

Breaking in: A background in agriculture, economics, development studies, natural resource management can be an advantage.

Job titles: Food Security Expert; Livelihood Expert; Agro-Enterprise Development Manager/Coordinator/Officer; Value Chain Development Expert

Find out more:

Gordon Conway (2012) *One Billion Hungry: Can We Feed the World?* New York: Cornell University Press.

Brian Gardner (2013) *Global Food Futures: Feeding the World in 2050*, London: Bloomsbury Academic.

Raj Patel (2013) *Stuffed and Starved: From Farm to Fork the Hidden Battle for the World Food System*, London: Portobello Books.

See also: Nutrition; Farming and agribusiness; Livelihoods; Economic development; Value chain analysis

---

I did a Master's in Development Studies at the National University of Rwanda and another at the International Leadership University in Kenya. I started working in community health, where I witnessed many avoidable deaths in children and adults due to hunger and food shortage. This sparked my interest in food security issues. In 2006, I became Area Coordinator with African Enterprise in one of the poorest provinces in Rwanda. Here the major challenge was soil degradation and acidity, which required innovative and creative solutions to enable the communities to produce at least enough food for their families to survive. We achieved this by introducing integrated farming techniques.

In 2008, I joined a conservation programme as Livelihood Enterprise Development Manager for CARE International, based in Uganda. Here we helped communities near the protected park areas to find alternative sources of income to move away from animal hunting and the destruction of the protected areas. I started looking at food security – not only as a way of having enough food but also being able to meet other needs from your harvest. We conducted value chain analyses and facilitating market linkages between producers and buyers, and we helped the communities raise income from their produce. I later worked as Food Security Specialist for Christian Aid, focusing in East and Central Africa. Here I saw how discouraging it is for farmers to produce a surplus, but not have the post-harvesting techniques and infrastructures, market access and storage to utilise this. This can easily transfer into a food insecure situation.

Increasingly I am focusing on accessibility – when food is affordable and there is good infrastructure to the markets for sale and purchase. I now work as Livelihood and Food Security Advisor with GOAL, mainly focusing on the Sahel region, which is very interesting. When I started working in Niger, I was expecting to apply some of the techniques I had successfully used in East Africa, but the context was so different I had to adapt my thinking and be more innovative to address the specific challenges of the Sahel. Here I am responsible for developing and leading the implementation of the food security strategy, conducting needs assessments, ensuring that humanitarian needs are addressed on an ongoing basis, as well as providing technical guidance to field-based teams and ensuring that high-quality activities are implemented in a timely manner.

I like the food security sector as it allows you to understand the various dynamics and interplay of development actors. It is a multidisciplinary sector, and I have learned almost everything on the job, from food production to food marketing, business planning, private sector engagement, policy analysis as well as advocacy. If you are looking to work in the food security industry, make sure you have a good development background, mainly in agricultural, economist or natural resource management. A good understanding of cooperative development, especially in developing countries, is very important.

Charles Karangwa (Rwanda) Burundi, Zimbabwe,
Uganda, Mali, Niger and Burkina Faso

Note

1 The World Food Summit of 1996.

FUNDRAISING

Money makes the world go around, and the fundraiser's role is to bring in the funding, enabling an organisation to carry out its work. Fundraising efforts can target the general public, trust and foundations or be from institutional donors that require more formal applications with detailed proposal and budgets. Each donor requires a unique approach. Larger organisations often have a team of specialised fundraisers, defined by their target group: community, corporate, legacy, major gifts, trusts and events. In smaller charities, a single fundraiser might work across all of these areas, targeting a range of different funding sources. INGOs generally focus their fundraising efforts in resource-rich countries, although there are increasing funding streams available in LMIC.

The work: Fundraising is an outward-facing job where you need to have a good knowledge of the external funding environment and what other organisations are doing but at the same time know your organisation's programmes inside out. Your exact work will depend on what type of fundraiser you are but could be anything from writing grant proposals (see also Proposal writing), planning a fundraising campaign to targeting individual donors to support a cause. Building relationships with your donors is important: understanding their priorities and exploring new fundraising opportunities. Moving up the ranks you might also be involved in designing an organisation's fundraising strategy, setting targets and managing a team of fundraisers. Evidence that you can 'bring in the money' is an essential characteristic for more senior management positions within NGOs, so fundraising can be a useful stepping stone to these.

The employers: The fundraisers' role is critical to the growth and survival of any NGO, consultancy firm or research institute or multilateral agency. Without a stable income, especially in these times of crisis, good fundraisers are in demand. NGOs of any shape and size will employ fundraisers; other not-for-profits will employ proposal writers or business developers rather than general fundraisers.

Breaking in: Evidence of successful fundraising is the biggest asset. It is mostly learning by doing, and many smaller NGOs will gladly take on volunteer fundraisers. The more you have managed to raise – from a range of sources – the better. Some undergraduate courses such as Marketing, Media or Business may give a slight advantage but are not essential. Knowledge of development and an understanding of your organisation's goals is vital so you *sell* your *product*. A few years in, gain a professional qualification to distinguish yourself from your peers. Becoming a member of a professional association such as the Institute of Fundraising (UK) will also be an advantage.

Job titles: Major Gifts/Individual Donor/Trust and Foundations Fundraiser; Grant/ Technical Writer; Director of Development

Further information: The Institute of Fundraising (UK), www.institute-of-fundraising. org.uk

Penelope Cagney and Bernard Ross (2013) *Global Fundraising: How the World is Changing the Rules of Philanthropy*, Hoboken, NJ: John Wiley & Sons.

Roy C. Jones and Andrew Olsen (2013) *Rainmaking: The Fundraiser's Guide to Landing Big Gifts*, CreateSpace Independent Publishing platform.

See also: Proposal writing

I studied Politics and Social and Economic History at Edinburgh as my undergraduate degree. I wanted to go to do a Master's so I wrote to a few educational foundations to seek funding. I was really astonished when I had a positive response from three separate foundations, almost covering my tuition fees. This made me feel I might possibly have a bit of a talent for fundraising. After my Master's (in Violence, Conflict and Development at SOAS) I decided to go for fundraising jobs as it seemed like a universally in-demand and transferable skill that could be useful both in the UK and in developing countries.

Fundraising is relatively easy to get into as you don't need any previous experience for many starter jobs and it's also relatively quick to move up the ranks. Head of Fundraising/Development jobs – should you stick it out that far – can pay pretty well. It is also an area where there are always jobs (just look at the number of fundraising jobs on *CharityJob* compared to any other skills area).

I started off as a fundraiser for some smaller UK charities, before moving to an international one. I was lucky in that one of my employers paid for me to complete a fundraising training course with Fundraising Training Ltd. I strongly recommend undergoing training early on in your career and establishing a contacts network.

I really wanted to work overseas in a developing country and was excited to find that fundraising was one of the skills that VSO recruited for. I felt this was a great way to getting international experience and also of having a chance to pass on my skills in a capacity building role. So I spent two happy years in Uganda.

I then became Fundraising, Partnerships and Communications Manager for Restless Development Sierra Leone. It's quite a big remit. I'm focusing on raising funds for the Sierra Leone programme both from in-country and overseas funders – supported by a small team based in London – while also supporting the programme's team to identify and approach non-funding strategic partners (government, other NGOs etc.) for programmatic collaboration. In the meantime, I am responsible for all external communications via web, newsletter, publications etc. There is no typical day, which is one of the great things about the job.

I enjoy having to have this bird's-eye view of things. Depending what area of fundraising you're in, there may be lots of face-to-face interaction with donors or it may be more of a behind the desk writing job. I personally like a bit of a blend of the two. On the flipside, it's hard work and involves a lot of disappointment and setbacks (it's quite normal that 80–90 per cent of your approaches/applications will be unsuccessful). However, when the money does come in, you're all of a sudden the hero of the hour! Even if it only lasts an hour . . .

Thea Lacey (UK) Sierra Leone, Uganda

GENDER

Gender roles differ greatly in lower income countries, where social and economic indicators consistently show that women bear the brunt of hardship in poor communities. Women are often found at the front of major global issues – food, water and energy security, population growth and climate change. Yet, unfortunately, the majority of foreign aid and development finance pursues social and economic outcomes that are blind to gender considerations. This therefore perpetuates the deep-rooted cultural barriers and discrimination that often run in parallel with poverty. Gender mainstreaming is a strategy to approach gender equality and avoid gender-blind policies and programmes. Gender-sensitive programming has become an imperative for many governments, donors and organisations alike, who are increasingly realising that development outcomes can be significantly increased when the different roles of women and men are considered at every stage of the policy or programme cycle.

The work: You will need to understand the gender relationships within the region and assess how the needs of men and women differ in each particular context as you will be helping to design the intervention accordingly for each development programme. As a Gender Advisor you may be working to assess programme design and integrate gender into all phases of the programme cycle. You may be researching any gender gaps, asking why they exist and how they can be overcome. Even if an organisation reports a 50:50 membership of men and women, what is the reality on the ground? Are women participating, benefiting and making the decisions equally? At the policy level you might be involved in developing gender strategies, or training an organisation to be gender-sensitive in their programme and not reinforcing gender differences. At the community level you might be working on women's empowerment (including sexual reproductive rights, for example) and promote gender equity in access to programme inputs, outputs and outcomes. At an evaluation level you could be reviewing sex-disaggregated data and understanding any challenges.

The employers: Some organisations such as UN Women (UNIFEM) and NGOs such as Women for Women target this specific group. But as gender mainstreaming becomes the norm within projects, there is increasing demand for gender experts within organisations to promote gender-sensitive programming. There are also many opportunities for consultants in the field to come in and deliver particular guidance, training or support in and reviewing of policies.

Breaking in: An advanced degree in sociology, anthropology or gender relations is advised. There are now many Master's relating to gender and development. Some work experience working on an organisation's gender policy or mainstreaming gender issues into programming will help you to move up the ladder.

Job titles: Gender Advisor; Gender Specialist; Gender Mainstreaming Officer

Find out more:

Janet Momsen (2009) *Gender and Development* (Routledge Perspectives on Development), New York: Routledge.
N. Visvanathann, L. Duggan, L. Nisonoff and N. Wiegersma (eds) (2001) *The Women, Gender and Development Reader*, London: Zed Books.

Armed with my Master's degree in Social Psychology, I joined government service. I started my career at the national level, working with the Philippine Department of Labor and Employment (DOLE). Here I initiated the review of policies and programmes related to work and training opportunities for women in the Philippines and those in Southeast Asia. Little did I know that this would lay the ground for my entry into international development work.

In 1988 I entered international development practice, in the technical staff of the ILO at Bangkok (Thailand) as the Chief Technical Advisor on ILO-DANIDA Sub-regional Project on Rural Women Workers. I was here until 1996. We were working towards strategies for the social protection of home-based workers (mostly women) who were usually not covered by social security laws and traditional social protection schemes. This network helped in the advocacy for the passage of the ILO Convention 177 for the protection of homeworkers. Note that this was against the backdrop of an evolving development scenario. Towards the 1990s, women as a vulnerable group gained attention through the Women in Development (WID) movement. Later the WID movement evolved to be known as Gender and Development (GAD), which adopted the gender mainstreaming approach. Since then there has been an upsurge in gender-sensitive programming and making governance and development gender responsive.

I then returned to my home country as head of the national skills development authority. On completion of this I was recruited as Regional Programme Director of the UNIFEM East and Southeast Asia Office from 2003 to 2005. Here I oversaw the various programmes and projects in the countries of the Southeast Asian region. Some examples of projects were a regional campaign to combat violence against women or the formulation of a gender and development programme in Mongolia. I am now an independent consultant undertaking assignments – mostly gender audits – for a number of UN agencies as well as other multilateral and bilateral agencies. I have also provided technical advice in the preparation of a manual on gender-responsive value chain analysis.

One of the most interesting projects that I have worked on was the promotion of women in politics in East Timor where we encouraged political parties to have a list of female candidates in their ticket. The results have been very positive with East Timor increasing the number of women in their Parliament. What motivates me in my work is my deep-seated and inherent interest in human development and how human potential can be harnessed for the greater good.

My advice to anyone wanting to work in gender issues is to broaden their networks to include civil society, people's organisations, as well as the private sector and intergovernmental agencies at regional and global levels. Stay connected and be sensitive to the emerging discourses and dynamics in development.

Lucita Lazo (Philippines) Cambodia, Vietnam,
Lao PDR, Indonesia, Mongolia, Thailand

GOVERNANCE

Good governance is a crucial component for sustainable development and underpins states that are democratic, accountable, transparent and bound by the rule of law. On the flip side, *bad governance* is associated with low rates of economic growth, corruption, poverty, insecurity and human rights abuses, all of which limit development and aid effectiveness. Since independence, many countries have struggled to develop a strong administration and lead their countries effectively. A successful administration needs to be flexible, and have the ability to reinvent itself continually to meet new challenges – critical to a country's ongoing success. Decentralisation from national to local institutions is one focus, ensuring services are delivered to the people. The discussions on budget transparency and open data have also been gaining momentum over recent years.

The work: From providing technical support and building the capacity of national and local institutions to be more effective, to working on legislative issues or research and policy advice. You might be working towards more consultative and inclusive development of policies, advising the local administration on implementing decentralisation strategies. You could also be advising the government on electoral procedures or doing election monitoring. Citizen participation is also a hot topic, ensuring participation in political processes, although this often happens more in rhetoric than practice. The work may be in fragile, post-conflict or transitional states.

The employers: Mostly bilateral and multilateral organisations, such as USAID, JICA, DFID, BTC, GIZ or WB. Some NGOs work in this area also, such as IREX. The Africa Governance Initiative (AGI) spearheaded by Tony Blair, also works in this area.

Breaking in: An academic background in a field such as public administration, law, political science, governance and development, economics, public finance or international relations will be an advantage. The IDS (Brighton, UK) for example runs a course on Governance and Development.

Job titles: Governance Advisor; Decentralisation Expert; Public Finance Advisor

Find out more:

Daren Acemoglu and James Robinson (2012) *Why Nations Fail: the Origins of Power, Prosperity and Poverty*, London: Profile Books.
www.governanceanddevelopment.com/

I studied Economics with a specialisation in Public Finance. But I came to development cooperation quite coincidentally. I saw a job advertised in Bosnia – a country I was interested in – with the German Technical Cooperation (GTZ, now GIZ). The job changed scope and I was invited to an open assessment centre. I was successful and was offered a two-year young professionals programme in the public finance stream, starting in Nicaragua. I weighed up the options: the Caribbean or a PhD on Financial Markets? I opted for the former and spent one year in Nicaragua, three months in GTZ headquarters in Germany, three months in Paris with the OECD and the remaining time in Ghana.

The experience gave me a broad overview of development cooperation, specifically in the field of taxation, and strengthening countries' own revenue to reduce donor dependency. After the scheme, I then took up a post to work for the German Development Service (DED – which now also forms part of GIZ) in Arusha (Tanzania) as financial advisor. Here I learned how local institutions really function, as I was placed directly in the Municipal Office to support the strengthening of local revenue generation. We worked on tax collection and introduced a trading tax. The project was successful and almost doubled the revenue of the municipality within two years, consequently featuring in national newspapers as a model. This has later gone on to form a component of a wider GIZ programme.

From here I went on to Benin, also with DED, working on decentralisation and was in charge of seven development advisors based in district offices. I just stayed for a year and then moved to Rwanda as Fiscal Decentralisation Advisor. The interesting difference here is that in Benin you rather had to push for something to happen, but in Rwanda I found that partners were pushing me: a nice change. There was a big opportunity to influence laws and policies. Among others we have been supporting the development of a new legal framework for local revenues and the elaboration of the decentralisation strategy. As an advisor we are not here to do people's jobs for them, or conversely, passively give technical advice. Instead it involves bringing people on board, helping them to understand and then develop and implement activities and strategies together. The challenge with decentralisation is that development cooperation projects often focus on national institutions, but how do you strengthen the services down where they are delivered to the people? Governance is not only at the national level but down to the level of the people.

On the plus side you have a connection with people, and jointly achieve objectives. On the downside you sometimes have to settle for second best. I will soon be leaving my position and heading back to the headquarters in Germany, an important step for my professional development before I take other international postings. The challenge has been recruiting a successor. It's difficult to find someone with the right mix of technical and advisory skills, especially in the field of Public Financial Management which is an area that has been growing in focus, with donors' emphasis on financial support.

David Lahl (Germany) Nicaragua, Tanzania, Benin, Rwanda

HANDICRAFTS AND DESIGN

The handicraft sector is becoming an increasingly important sub-industry in many developing countries, employing a considerable number of people. It is one of the principal sources of *livelihood* available to people living in precarious circumstances – as production is often conducted at household level – ideal for single mothers, small farmholders, war victims, and people with disabilities. The challenge is that artisan enterprises are highly vulnerable, as successfully marketing and selling hand-crafts is complex and difficult. It requires strong business skills, market linkages and the ability to develop new products based upon their customers' interests. Local markets are important, but with good links to the export markets, there is much potential for handicrafts to contribute to the local economy, especially in rural areas.

The work: Working in product development you will be harnessing and translating indigenous skills to modern tastes, creating handmade designs that meet the requirements of the export and tourist markets. This will involve the design and costing of new lines and products, for example. You might also be in charge of the selection of materials and colours to develop unique and high-quality products. In marketing you will be involved in developing advertising tools for the organisation, identifying clients and linking to markets. The Internet is essential for working with export buyers and an important tool for reaching new markets that many of the artisans themselves do not have access to, which highlights the importance of international support. As in any business, market research and understanding the demands of the customers are highly important. In all roles there will likely be a strong element of capacity building and training local people in how to run their business more effectively.

The employers: Some NGOs work to equip local artisans and cooperatives with the skills required to produce the right products for the market and train them in costing and pricing, accounting, communication, promotional activities that reach their target, etc. Aid to Artisans, www.aidtoartisans.org is one such organisation. Other social businesses work with artisans directly training them to make their products and buying directly from them for resale on the international market. Business, market-driven models are the best way to ensure success. Other international organisations, such as the ILO, are involved in policy and trade.

Breaking in: From business training, product innovation and marketing the work can be very broad, reflecting the various routes in. A design background, adult education, business training, buying or marketing will give you relevant and in-demand skills. Some Peace Corps volunteers are also placed with artisan cooperatives to enhance market linkages, a good way to gain experience.

Job titles: Designer; Production Manager; Marketing Manager

Find out more:

www.ilo.org/public/english/region/asro/bangkok/library/download/pub06–16.pdf

See also: Livelihoods

I studied a BA in Textile Design with a specialism in embroidery, because it was what I loved. After graduating, a family friend connected me with a Tanzanian organisation, WomenCraft. The work they did looked interesting so I contacted them. They agreed to take me on as a volunteer, so I first had to work at home to raise some money.

WomenCraft works with basket weaving, for export and for sale on the local market. I started by helping out with basic tasks such as rearranging documents and choosing fabric at the market to weave into the baskets. A lot of my time was also spent adapting designs from Western designers, who had contributed beautiful designs, but hadn't had the time to work in the country and develop the traditional techniques in a way that would make their designs work. We also held a design competition with the artisans: it was interesting to see how taste differed between judges, with Westerners and locals often divided, which indicates why, especially for the export and tourist market, having international designers is so important. I extended by another six months and took on more responsibility, started to do more design and introduce my own range, a new weaving technique and a series of bags.

After a year of volunteering I realised that this work was my calling and began to look for a more stable paid position. I began by looking on all the job sites, trying to find opportunities and organisations with a design component. I now work for All Across Africa. The organisation advertised for an in-country manager on Idealist.org. I applied, but highlighted that my interest was in design. I was pleased to find that they were willing to discuss my fit within the organisation. My current job title is Production Director and Designer. Working in a niche area such as design makes you a valuable asset to a handicraft-based NGO; it can be very challenging for people to find permanent design staff who are willing to work in-country as most designers want to be at the cutting edge – close to what is trending – not working in remote rural areas in far-off countries.

I design baskets, paper beads and bags exclusively for the export market. My days are never the same and I may be searching for materials online or in local markets, developing patterns, teaching a new technique, training in product specifications or doing quality control. We work with local cooperatives, predominantly made up of women. We also offer them support in managing their finances, setting up a bank account, book-keeping and reconciliation training. The better they run their business the better we can run ours and we can both grow.

What I like most about the work is that by connecting artisans to new markets we can revive, and keep alive, traditional techniques, adapting them to make something functional in the modern world. It is also mutual learning and I enjoy sitting with the women to learn from them. We may have trouble communicating with words, but there is a shared language in the craft. When developing new designs its very participative, with what they feel they can do, and what we need for our clients. Many of the women are struggling with poverty and we provide them with a market for their products and a livelihood for their families that can be transformative.

Eleanor Dart (UK) Tanzania, Rwanda

HEALTH PROFESSIONAL

Worldwide there is a global shortage of health workers, and 57 countries around the world have been identified as having a critical shortage – 36 of these are in sub-Saharan Africa. There is a need to build up and develop local services, as many specialisms and a continuous supply of professionals simply do not exist, often leading to medical malpractice and unnecessary deaths. Whether you are a midwife, nurse, doctor, lab technician, phlebotomist or health services management, a health profession travels well. In a humanitarian crisis health care is one of the most immediate needs, as local services are often unable to respond to the rapid concentration of injured or traumatised people, requiring external health professionals to be brought in. During reconstruction efforts and in developing country contexts there is also a high demand for health professionals as most countries do not train enough to meet their needs.

The work: Some positions may be short term, such as in the immediate aftermath of a humanitarian emergency, where there will be a lot of hands-on work, 'putting out the fire'. After the situation stabilises – and in development contexts – some practical work might still be needed, but the emphasis will be on *capacity building* and thus often requires longer-term roles. Here you will have a greater role in training others, managing projects, policy work, developing local services and contributing to the next generation of health professionals. There are likely to be many frustrations too, as the lack of equipment, skills and treatment options means that you can't practice, or train others to practice, as you have been trained to do.

The employers: Some of the main organisations that recruit health professionals for humanitarian work are Médecins sans Frontières (MSF), Doctors of the World, Merlin, The Red Cross (ICRS and IFRC movements). Many will recruit in development contexts including VSO, UNV, Healthserve. There are opportunities for both full-time careers in international health or it is also possible to combine it with your work back at home. *Links* may be facilitated via UK health institutions through organisations such as Tropical Health Education Trust (THET). Some organisations such as Operation Smile, Orbis, Mercy Ships recruit professionals for as little as two weeks.

Breaking in: Explore whether you can include international placements as part of your training requirements. An elective educational placement is a good time to go overseas and gain as much additional exposure to working in a developing country context. Build up at least three years post-qualification experience. If you want to work in humanitarian emergencies a first step might be to undertake a short introductory course such as RedR UK's 'So you think you want to be a relief worker'. Individual organisations also run training courses for people wanting to work with them such as Medair's eight-day Relief and Rehabilitation Orientation Course (ROC).

Find out more:

Maïa Gedde, Susana Edjang and Kate Madeville (2011) *Working in International Health*, Oxford: Oxford University Press.

See also: Public health; HIV, AIDS, tuberculosis and malaria; Maternal, child and reproductive health

---

I was born deep in the jungles of West Papua to my missionary (doctor and nurse) parents, which may have sparked my interested in international health. Growing up I was struck by injustice. After graduating in Medicine from Aberdeen University (Scotland) and receiving specialist accreditation in palliative medicine I worked for Grampian NHS Trust as a senior consultant and honorary senior lecturer for ten years. During this time I sought and took opportunities to develop my skills and knowledge about international palliative care. This included short training visits to Belarus, Ukraine and India while also running a busy regional palliative care service and setting up undergraduate and postgraduate training, reviewing the medical school curriculum, senior management and national advisory roles as well as an integrated clinical service. With this senior experience it is essential for me to now offer these roles internationally.

With the challenges of combining international work with responsibilities back home, I decided to leave the National Health Service (NHS) in the UK and make a full-time career internationally. My motivations are justice and shared humanity. The shocking reality that four out of five people worldwide lack access to pain relief medicines and that 3.6 million die with untreated pain is an international scandal. Holistic care can also be given without high-tech medicine and allows us to work together as health care workers and the wider community to support those in need.

Increasingly I learned the important role international colleagues can have in working in partnership, building capacity, offering senior mentorship and advocacy and at times supporting direct projects and services.

But as I started to research I realised that there was a distinct lack of organisations working in the field of palliative care so I helped found a Scottish NGO (Cairdeas International Palliative Care Trust) which has a vision of ensuring palliative care for all. I then moved to India to be involved in the development of postgraduate curriculums and offer mentorship, always in partnership with Indian colleagues and organisations.

One of my main interests is in health systems strengthening and integrating palliative care and so it was an honour to be invited by Makerere University (Uganda) to set up one of the first academic palliative care units in sub-Saharan Africa. We collaborate with the Global Health Academy, University of Edinburgh, run an exciting health systems strengthening project in East and Central Africa as well as support the delivery of postgraduate programmes. I also spend at least a month every year in north India where palliative care is very sparse. You can imagine my challenging and varied life; so far this year (June) I have visited ten countries.

There are many, many frustrations and challenges, particularly working in resource-limited settings where systems are not efficient and corruption is a way of life. Building capacity means allowing others to make mistakes and do things in their own way yet that partnership is the rich place of mutual learning. If you want to work in this field, start with offering time on short visits or volunteering in order to learn. Ensure you have something to offer, and be well trained in your own area of expertise including experience at senior levels such as service development, research and training. You might also consider a Master's in Education.

Mhoira Leng (Scotland) India, Uganda, Nigeria, Zambia, Kenya

HIV, AIDS, TUBERCULOSIS AND MALARIA

These devastating diseases predominantly affect the poorest in low- and middle-income countries. While those working in this area fit into the public health or health professional categories, it was felt that these infectious diseases merited a section of their own. In the last decade, substantial amounts of funding have been channelled specifically towards them, enabling progress. Today for example, it is possible to prevent mother-to-child transmission of HIV: an important stride towards an AIDS free generation and the elimination of the disease. Yet, rolling this treatment out remains a challenge and 900 babies continue to be born every day with HIV. Inadequate health systems remain one of the main obstacles to scaling up interventions and securing better health outcomes. After AIDS, TB is the second leading cause of death from an infectious disease worldwide. TB is also known as a disease of poverty, affecting mainly young adults in their most productive years. The vast majority of deaths from TB – over 95 per cent – are in the developing world. Malaria is endemic in 106 countries.[1] In 2010 there were an estimated 655,000 malaria deaths – of which 91 per cent were in the African region. Approximately 86 per cent of malaria deaths globally were children under five years of age.

The work: You might be working towards supporting the development of a country's sustainable HIV and AIDS response by strengthening their health systems. Or you could be doing surveillance work, monitoring uptake of initiatives, working in voluntary testing and counselling centres and building the capacity of health organisations, training health staff. The work could also involve rolling out a programme of insecticide-treated bed nets to prevent malaria or overseeing the management of a directly observed trreatment, short course (DOTS) programme, at the heart of the Stop TB Strategy. Procurement and supply chain management are also essential, ensuring the right drugs reach the right people at the right time.

But the work could also have a human rights angle, ensuring that those infected or affected are not discriminated against, and receive equitable treatment.

The employers: The Global Fund to Fight AIDS, Tuberculosis and Malaria was established in 2002 and is the main multilateral funder of these infectious diseases. The WHO, UNAIDS, the GAVI alliance (promoting immunisation) and the roll-back malaria consortium also work in this area, as do a host of NGOs.

Breaking in: A background in health sciences, public health or counselling will be an advantage but there are also many roles that you can do, such as project coordination or management, that do not require these skills, but rather a background and experience of the field and programmes.

Job titles: HIV and AIDS Specialist

Find out more:

Chinua Akukwe (2006) *Don't Let them Die: HIV/AIDS, TB, Malaria and the Healthcare Crisis in Africa*, London: Adonis & Abbey.

Each organisation has their own publications: see for example those of Roll Back Malaria www.rollbackmalaria.org/multimedia/partnershippublications.html

See also: Health professional; Public health

A family situation made me acutely aware of the challenges of being sick and poor. This got me interested in the intersection between health and politics, but it was at a time when Public Health wasn't a popular field of study for those without medical training. During an internship at the Harvard School of Public Health I found out about the *social determinants of health* and was fascinated (and shocked) by the evidence on how social factors can influence a person's health status. This got me interested in why poor people have worse health and what can be done about it. I was born in Trinidad so I have always had an international outlook and an interest in poorer countries. After my first degree in Biology I went on to pursue a Master's in Population and International Health. I decided to specialise in HIV and AIDS as it is an area of health that is extremely cross-cutting, bringing together the health, politics, and cultural aspects of the society, using an international lens.

After I graduated, I joined John Snow Inc. (JSI) based in Boston where I worked mainly on maternal and child health projects. This job gave me the essential skills in project management to complement my public health background. I decided to move to the not-for-profit sector of an HIV fund based in New York. I am now the Director of Global Programs at International Treatment Preparedness Coalition (ITPC). The organisation provides treatment access support to communities in need through knowledge building, advocacy and coalition/movement building. Our programme work includes small grant-making to local organisations across 12 regions of the world. An example of our work includes supporting projects that work with people with disabilities who are denied surgery because they are HIV positive to enable them to access better care and support. We also do advocacy, and recently produced a *Missing the Target* (MTT) report in response to a WHO report showing many successes, but that was not truly representative of the reality on the ground. We also work on other global advocacy priorities including work on unfair trade practices that deter access to life-saving antiretroviral medicines.

My current job requires frequent travel to conduct trainings or to attend a technical meeting. For example, I recently helped facilitate a pre-implementation workshop in St Petersburg, Russia, for a grant round. We brought all our community-based grantees together to learn about reporting requirements, templates and monitoring and evaluation. The agenda included discussions on how to effectively track the number of pregnant women referred to antenatal clinics and how to use this data to reduce loss to follow up when looking at their clients.

For anyone considering work in international public health, specifically HIV, it is important to realise that health is broad and career paths are not usually very direct in this field. Learn from *all* your experiences and remember, knowing what you want to do is *just* as important as knowing what you *don't* want to do!

Solange Baptiste (Trinidad/USA) Ukraine, Pakistan, Kenya,Panama

---

Note

1   Accurate in 2010.

HUMAN RESOURCES (HR)

Recruiting the right people, with the relevant technical experience and inter-personal skills is crucial for the success of all humanitarian missions and development projects. As so much of this work is project based, mostly with short- and medium-term time frames, a lot of hiring takes place. Not only do the right people need to be put in post, but they need to be adequately supported and managed, to ensure that they are able to deliver their best. Organisations bidding for tenders or grants have a unique recruitment process that begins at the proposal stage before any funding has been secured – for this reason you may see some job adverts state 'funding pending' – even if you are recruited for the role there is no guarantee that the organisation will win the tender to implement the project. Conversely, humanitarian missions may require people immediately, the recruitment process will happen quickly and deployment will be rapid.

The work: Working in head office you may be responsible for a range of HR tasks – this could include recruitment, payroll, employee relations, induction and briefings, learning and development and exit interviews. You may either specialise in these areas or be expected to cover all of them – this will depend on the size of the organisation. Working in a field mission you are likely to be a generalist in all of these areas, being involved in the planning and design of required teams, ensuring the implementation of all HR policies and management guidelines and supporting managers to identify talent and lead their teams. Some HR roles at field level will also have an element of administration that will include travel, arranging visas, and office and guest house management. HR demands a wide skill set – being able to get on well with people, as well as being organised and able to think about the long/short term and organisational risk are all key.

The employers: All medium and large organisations working in the sector will have an HR function and HR will typically be a member of the management team.

Breaking in: Many employers now look for a qualification in HR and, more importantly, some experience in generalist HR roles will be required.

Job titles: HR Manager; HR Business Partner; HR Advisor; HR Officer; HR Coordinator; Recruiter

Find out more:

Chartered Institute of Personnel and Development (CIPD) (UK based) www. cipd.co.uk/
People In Aid – www.peopleinaid.org/

See also: Project management; Proposal writing

I completed an undergraduate degree in History and secured a personnel role in the public sector straight after university. I was fortunate enough to be financially supported to study part time for a Postgraduate Diploma in Personnel Management at the very early stages of my career, leading to membership of the CIPD. Subsequently I worked as a personnel manager for various NGOs, and then by chance I spotted an advertisement for a Regional HR Officer with Oxfam GB. I applied and was successful, first working from Khartoum, Sudan and covering the Horn of Africa region for three years, and later moving to the Middle East, Eastern Europe and CIS region for another four years as Regional HR Manager. In both these roles I provided HR support and management to the region across different countries, wrote and delivered the HR strategy and led on implementation of work plans. The HR work focused on change management, dealing with poor performance, reorganisation of team structures, management development, job design and a review of HR practice and terms and conditions. I also had a five-month secondment to Darfur to lead the HR of the emergency response programme in 2004.

After this I was the HR lead for Oxfam GB in the tsunami humanitarian response based in Aceh during 2005. Here I led the recruitment of 700 staff during the start-up of the response programme, all of whom required inductions and employment contracts. I also worked in Pakistan to support the earthquake response programme where I led on the reduction of staff teams and then closure of the earthquake programme. Following Oxfam, I joined Save the Children UK, as Head of Talent and Senior HR Business Partner. I have recently undertaken a three-month deployment to Save the Children Australia working as country HR Consultant in the Solomon Islands and am now the Director of HR at MAG (Mines Advisory Group).

HR work can be complex, rewarding, varied and full of surprises. It's an opportunity to work closely with the organisation's leaders, to understand the issues and challenges and to make a valuable contribution to move things forward. A career in HR offers many opportunities to work in a wide range of contexts in either a specialist or generalist role. My colleagues and peers come from many different backgrounds: some have worked in the corporate sector, some have come from project management backgrounds and some have stayed in the same organisation and worked in different functional teams throughout their career. All have gained valuable experience at different levels and most have qualified in HR management.

Interviewers like to see that you may take a risk on an opportunity, are prepared to gain experience and to learn at a lower level before expecting to reach the more senior positions. Taking an interest in the business of HR should not be a stand-alone function. Apply for roles that might not be exactly what you are looking for but would give you valuable experience.

Samantha Wakefield (UK), Sudan, Kenya,
Pakistan, Indonesia, Solomon Islands

HUMAN RIGHTS

International human rights law lays down obligations of governments to act in certain ways or to refrain from certain acts, in order to protect and promote the rights and fundamental freedoms of people within their territory. However, despite international commitments, governments around the world continue to violate those rights and fail to fulfil their obligations. Human rights are increasingly seen as a vital component of development: by considering issues such as health, housing and education as human rights concerns, people are empowered to ask for those rights to be respected within a clear legal framework. Many human rights violations may involve discrimination in access to services on grounds such as sex, race, ethnicity and disability. But it can also involve, for example, harassment, mistreatment and intimidation of human rights defenders and activists (such as land rights activists). Violence, impunity, intimidation and failure to respect the rule of law create an environment in which development, especially equal development, can only be hindered.

The work: By specialising in this area you could work as a researcher, policymaker, advocate, community trainer or a campaigner. You may be undertaking independent human rights investigations or analysing laws and policies for their compatibility with international human rights standards (in theory and in practice). Working within government institutions you could be advising on legal reform or the use of human-rights based approaches in national and local programmes, or empowering and supporting community groups and activists in using human rights as a tool to ask for changes in policies and practices. There are many different substantive areas you could specialise in such as refugee protection, gender, child rights, housing/land rights or justice/rule of law.

The employers: Many traditional 'aid' organisations are increasingly using human rights approaches in their work and many embassies and government agencies have dedicated human rights officers. Examples of some organisations include UN agencies such as Office of the High Commissioner of Human Rights (OHCHR) and UNHCR and UNDP; Amnesty International; Human Rights Watch; One World; Derechos Human Rights; Fédération Internationale des Ligues des Droits de l'Homme (FIDH); The International Bar Association's Human Rights Institute; Interights; Islamic Human Rights; Redress; Survival; Council of Europe.

Breaking in: There are no clear career entry points, routes and structure for working in human rights. The field is increasingly popular and highly competitive to get into. Many practitioners have a legal background, although researchers may also have a background in journalism or the social sciences. Employers will look for a postgraduate qualification in human rights. Many organisations offer internships and this will help you to get a foot in the door, although they are often unpaid. It can be useful to find your niche in the field, becoming an expert in a particular area such as women's rights or refugee protection.

Potential job titles: Human Rights Advisor/Officer; Human Rights Campaigner; Human Rights Researcher; Children's/Women's Rights Specialist

Find out more:

Paul Gready and Wouter Vandenhole (2013) *Human Rights and Development in the New Millennium: Towards a Theory of Change*, London and New York: Routledge.
www.hri.ca
www.un.org/en/rights/

See also: Protection; Law and development; Disability

I've always wanted to work in human rights and studied Law (undergraduate in the UK and Master's in Law at New York University (NYU)) with the intention of making a career in this area. I have family from South Africa so grew up with the ever-present background of apartheid, giving me a strong sense of justice, and injustice. It's not an easy career to break into though and while I knew I didn't want to be a corporate lawyer, I had a difficult time when friends were transitioning from law school to law firm and I was waitressing and trying to figure out how to make my break into the human rights sector.

An unpaid internship or volunteer work seems the norm, although there are some opportunities for funded internships. NYU funded me to spend three months with the UN High Commissioner for Refugees after my Master's in Law (LLM), and there are bodies such as The Law Society in the UK that fund recent graduates to take human rights placements overseas. I think my first break was getting a six-month internship at the International Criminal Tribunal for the Former Yugoslavia in the

Hague after my Bachelor's in Law (LLB), where I assisted the tribunal judges with legal research and case management.

My first proper (paid) job in human rights was with the UK Foreign and Commonwealth Office where I worked for two years as an advisor on issues relating to justice and detention for British Nationals overseas. It was mainly a UK-based role but I was able to travel to the US, Thailand, Malaysia and China to train British diplomats on human rights and visit prisons, including death row. I have since worked on human rights in the UK Parliament and then as a Policy Advisor to the UK section of Amnesty International where a large part of my work involves analysing UK foreign policy and how it contributes to (or hinders), the improvement of human rights situations around the world.

I decided that I wanted experience working in a 'field' environment so I took a sabbatical to take up a VSO placement in East Africa. I'm helping organisations to examine the extent to which people with disabilities are facing discrimination in access to basic services. We are conducting research and analysis that will result in a report to the UN Committee on the Rights of Persons with Disabilities.

I like the work because the successes make a real difference. I've also had a lot of variety in my career, from research, writing policy papers to facilitating trainings, speaking at events and acting as a media spokesperson. But, equally, it can be frustratingly slow – you can plug away at an issue for years and nothing changes. In some countries human rights can be controversial so you have to be careful what you say. To break in, gain some experience, if not through an internship, try volunteering with refugees or a community organisation in your home country – you'd be surprised how many issues have rights at their heart.

Tara Lyle (UK) East Africa

HUMANITARIAN RESPONSE MANAGEMENT

The humanitarian response manager coordinates and implements their agency's effective and efficient response to humanitarian crises. They oversee the assistance provided to the affected communities during a humanitarian intervention as well as deployment, staff safety, information and risk analysis. During recent years the humanitarian community has initiated a number of interagency initiatives to improve accountability, quality and performance in humanitarian action and the humanitarian response manager ensures adherence to these standards as well as representation in clusters, task forces and regional teams. Four of the most widely known initiatives are the Active Learning Network for Accountability and Performance in Humanitarian Action (ALNAP), Humanitarian Accountability Partnership (HAP), People in Aid and the Sphere Project.

The work: It is often fast-paced and unpredictable although you won't always be based out in the field, often coordinating the response and appropriate deployment of staff from the head office. Your work will tie in to the humanitarian programme cycle: a rapid assessment followed by planning appropriate strategic interventions, recruiting the right people and monitoring and evaluating the impact of the work. You will bring together a team of appropriate professionals, who may focus on logistics, public health, food security, gender, protection, education or WASH. You will need to ensure implementation is in line with international humanitarian standards and best practices; that the scope, budget, quality and timescale, and grant and sub-grant obligations are adhered to; and manage and develop staff to ensure quality programme implementation and national capacity development. When not working directly on a crisis, you may be involved in *programme development*, identifying possible future humanitarian needs and planning appropriate strategic intervention.

The employers: NGOs are among the first responders in providing assistance as they often have strong links with affected communities. The United Nations (UN) is also a key player in humanitarian response with agencies such as the WFP, the WHO, UNICEF and the UNHCR. Coordination of global humanitarian response efforts is through the United Nations Office for the Coordination of Humanitarian Affairs (UNOCHA).

Breaking in: You will need to have knowledge of humanitarian operations and experience of conflict or immediate post-conflict environments. One training programme is the Humanitarian Leadership Development Project (HLDP), a 12-month programme focused on bringing new leadership talent into the humanitarian sector run by the Consortium of British Humanitarian Agencies (CBHA). There are also other humanitarian leadership training programmes and various Master's courses that can be done as full- or part-time courses as well as shorter taster courses.

Potential job titles: Humanitarian Programme Coordinator; Emergency Response Programme Manager

Find out more:

Martin Christopher and Peter Tatham (2014) *Humanitarian Logistics: Meeting the Challenge of Preparing for and Responding to Disasters*. London: Kogan Page.
Alessandra Cozzolino (2012) *Humanitarian Logistics: Cross-Sector Cooperation in Disaster Relief Management*. London/New York: SpringerBriefs in Business.

See also: HR; Logistics

Humanitarian response management is a vast, varied and fast-paced job. I didn't set out to become a humanitarian response manager at all, I studied English, but after a 'gap year' in India volunteering for development organisations I discovered that this was something I was far more interested in and that suited me perfectly.

The field is difficult to break into and while you are waiting, be patient and a bit creative if you don't get lucky. I was on my third or fourth development job on a local salary working overseas when I began studying for an Open University Master's in Development Management. The degree taught me lots of the skills of managing large-scale operations as well as quite a bit of development theory that I immediately applied in my work. Before I had finished the degree I had a full-time job in the field, working for Oxfam.

It took me two more years before Oxfam offered me my first overseas posting in a management position. I then worked in South Asia, East Asia and most recently in DRC, Ethiopia and Sudan. In my last job with Oxfam I was managing an overall country programme with a large humanitarian element – showing that there are places to develop still further applying those management and technical skills if you are interested.

A day's work depends on what comes onto your plate in the morning. It can range from ensuring you have the right security arrangements in place for the team, to attending coordination meetings and contributing your analysis from the programmes you manage to an overall country situation analysis. You could be reviewing a staff member's performance with them, sorting out logistics systems, planning a multi-sectoral programme for the next three years (in areas such as food security, food aid, nutrition, health, water and sanitation), in the car on your way to visit a project site or a newly affected emergency area, or giving a radio interview live on the air.

The work is high-paced and as a leader of a team, even in sometimes insecure and stressful environments, it is your job to set the direction, and to keep all of your staff safe, healthy, and continuously performing even under pressure. While that can be highly motivating, it's also important that you have your own hobbies or separate space to retreat to in order to rest, even if it's only to switch off and read a book for half an hour before sleeping. As the leader of the team, you take much of the praise, but also much of the blame if things do not work. It's important that you have support networks too – a line manager, even at a distance, could also be a good mentor, and having friends outside work (even from other NGOs or agencies) has often been a life-saver for me.

There are many different career trajectories to take from being a humanitarian response manager. I'm now working for the British Government as an advisor on humanitarian work.

Juliette Prodhan (UK), Sudan, DR Congo, East and South Asia

INFORMATION AND COMMUNICATION TECHNOLOGY (ICT)

Technology and development are inextricably linked, and we are increasingly seeing that ICT is a powerful means to change the lives of individuals in developing countries, and enhance their livelihoods. Fibre optic connections and high speed wireless connections (3G and 4G) are available in nearly all capital cities around the world and connectivity is growing, especially in Africa. Mobile money technologies based on SMS are extending financial services to everyone. Transactions worth 25 per cent of Kenya's GDP take place over the m-Pesa mobile money service that employs over 50,000 people. Mobile technologies are delivering health services and allowing agriculture to be more efficient. E-government and e-commerce is enhancing service delivery to all everywhere. Renewable energy technologies allow ICTs to be used beyond the reach of the electrical grid (off-grid technologies) and developing country entrepreneurs are seizing the opportunity by developing tools and applications to serve the underserved.

The work: From developing the technologies, to promoting roll out, advising on uptake and making ICTs more efficient and inclusive. You may be making recommendations about how ICTs can be used in various sectors, demonstrating to local decision makers the benefits of allowing the greatest access possible to ICTs (cost–benefit studies), while pointing out the disadvantages of not doing so. The sectors that ICTs impact are wide ranging and you will most likely be working with a particular sector focus: education, health, government, business, trade, climate action, the empowerment of women and marginalised groups, youth, local and community groups, service delivery in the public and private sectors, etc.

The employers: The World Bank Group institutions and the regional development banks, UN agencies such as UNDP and the UNOPS and bilateral aid agencies are important employers. Private foundations such as the Gates Foundation and the

Clinton Initiative also invest in ICTs, but usually as a mechanism for assisting with the delivery their focused assistance. Many NGOs such as Practical Action also work in this area. Private sector is important, particularly in the extractive industries, telecoms and mobile phone operations. Local innovation centres also have a role to play.

Breaking in: A Master's in an ICT-related field is a good start, especially if it focuses on a priority area that is just getting traction. For example, ICTs for sustainability and for climate and environmental action is a growing area of endeavour. For conventional ICT for development (ICT4D) training, the University of Manchester in the UK and the University of Hong Kong are strong. Several USA universities have ICT4D departments. Look for departments with as much geographical expertise and coverage as possible. Many NGOs also support the diffusion of ICTs so a route in can also be working with this type of organisation.

Job titles: ICT4d Strategist; E-governance Specialist; Knowledge Management Specialist; Expert in Mobile m-Agriculture, m-Health, m-Education; Market Analyst, Mobile Technologies for Development; ICT for Business Development

Find out more: A good resource is the website www.ictworks.org. You can also join the ICT4D on LinkedIn.

My background is in biological sciences with a Master's in Plant Physiology. I was interested in applied biology as a way of solving global food problems and wanted to work internationally. I got a job with the International Development Research Centre (IDRC) of Canada to set up an agroforestry information service in Kenya. It was in this role that I learned how to use ICTs for managing and sharing knowledge as I developed projects to help agricultural research institutions in developing countries manage and share information. This is where I learned the foundations of developing databases on PCs.

At the Earth Summit meeting in Rio in 1992, I linked up with a project to implement the conference agreements focused on the use of ICTs as tools for development. I helped NGOs, government agencies and others to learn about PCs, their application, email and eventually, the Internet. I started working as a consultant for the UNDP, which I continue today.

I decided that the only way to establish myself was to undertake as many missions as possible and learn about different development situations. As an example of my work, I was recently in Chad, developing a strategy to enhance financial inclusion, strengthen trade and integrate Chad in the regional and international trading system. I looked at the constraints to greater trade integration and proposed policies and projects to overcome these limitations and promote trade in Chadian agricultural and other products and services to the benefit of all citizens across the country. More recently, I have been advising the government of Swaziland on how they should implement their e-government strategy.

The ever-changing nature of the work is very stimulating as is the opportunity of continuously learning about ICT applications and development issues in general and their application. The main challenges affecting the use of ICTs remain government policies that constrain the opening up of business and ICT markets, and especially telecommunications markets, and allowing the greatest competition possible along with investment. The other focus is on developing the human skills to allow the technical and managerial transformation required to facilitate access to ICTs to take place.

If you want to work in this field, track global trends and try to understand where ICTs can be most useful in helping to resolve development challenges. Do this in the same way you would invest money: identify opportunities, assess risk and make the investment. Establish your presence and expertise using LinkedIn, Twitter and Facebook. Get to know the organisations, their work and the issues. Finally, and last but not least, stay in good physical shape, avoid politics and risky behaviour and drink lots of water!

Richard Labelle (Canada) Kenya, Sierra Leone,
Bangladesh, Egypt, Chad

KNOWLEDGE–POLICY INTERFACE

Knowledge-informed policymaking promotes a culture in which decision-making processes in – but not limited to – governments, regularly engage with knowledge when formulating, implementing and reviewing policy. This includes research, citizen input and practical experience. There are those who do this directly, by working with civil society or governmental organisations, and those who provide capacity development support and services to those who do this (such as representatives of donor agencies, international organisations and consultancy firms).

The work: You might be equipping policymakers with the tools and methods to be able to procure and use science cost-effectively; adding value to scientific research by ensuring its implications are well understood; and/or working with intermediary organisations that facilitate and convey policy debates and narratives. This in turn might involve undertaking research as well as providing tools and improving the skills of others in the following areas: decision-making processes, policy engagement, managing change processes, developing networks and partnerships, communicating knowledge/research and monitoring, learning and evaluation.

The employers: Organisations with an international development focus working in this area include international consultancy firms (such as McKinsey's), international NGOs (such as the Asia Foundation), most bilateral and multilateral organisations (such as DFID, WB, ADB and UNDP), universities (such as Oxford and Australian National University) as well as international development think tank/consultancies (including ODI, IDS, IIED and others).

Breaking in: No particular academic subject will necessarily give you a head-start – you might find people with degrees from astro-physics to development studies and zoology. Developing country experience is vital and so is practical experience trying to effect change in some way – whether it be as a 'knowledge broker' or an NGO practitioner, a civil servant trying to promote reform or as a researcher communicating the findings to policy audiences. A PhD is not essential to enter the sector, but can strengthen bids and funding proposals (if working for a consultancy).

Job titles: Research Officer; Research Fellow; Knowledge Broker; Knowledge Exchange Officer; Research Uptake Officer

Find out more:

H. Jones, N. Jones, L. Shaxson and D. Walker (2012) *Knowledge, Policy and Power in International Development: A Practical Guide*, Bristol: Policy Press.
www.ebpdn.org
www.impactandlearning.org/
www.knowledgebrokersforum.org/

See also: Research and academia; Policy

I undertook a Master's programme in Manufacturing Engineering and Management from the University of Durham. In my final year I applied to VSO to work as a teacher (back then – in 2002 – VSO recruited fresh graduates as teachers). I spent two years as a volunteer physics teacher in Kabompo, a town in the North-Western Province of Zambia. I had a wonderful time teaching, pursuing extra-curricular activities and doing my best to raise awareness of HIV and AIDS prevention among students and teachers alike. I also developed an understanding of the challenges faced by Zambia's education and health sectors. After my placement came to an end, I was asked to work at the VSO country office in Lusaka, which I gladly agreed to. It was here I learned the basics of 'NGO management', getting involved in strategic planning, fundraising, partnership development, project coordination and advocacy work.

After 15 months based in Lusaka I returned to London, UK to undertake a Master's in Development Studies at SOAS to help link my practical experiences with some of the broader theoretical and philosophical frameworks. During my studies I interned at Interact Worldwide – a sexual/reproductive health and HIV and AIDS NGO in the UK. Given my experience in Zambia and background studying manufacturing and management including global value chains, I tried to combine the two by writing a thesis on the costs, risks and benefits of Zambia's participation in the fresh vegetable value chain.

After my Master's came to an end, I managed to secure two part-time roles in London, a paid post at a health-focused NGO and an unpaid internship working with the ODI coordinating the Forum on the Future of Aid (FFA). After a few months I was offered a full-time position in my first role, but I decided to continue to hold down both part-time roles as the second was more in line with the work I wanted to do. Things changed in early 2008 when I successfully applied to ODI's Research and Policy in Development (RAPID) programme as a full-time Research Officer. Four years later I was promoted to Research Fellow.

My work involves undertaking proposal development, doing research, developing strategies and plans, facilitating workshops, producing toolkits, mentoring others, coordinating networks as well as monitoring, learning and evaluation. Some of the work I have been involved in includes a three-year study to understand the diversity of think tanks in Latin America, sub-Saharan Africa, South Asia and Southeast and East Asia; a three-year UNDP funded project to improve the capacity of the Vietnamese Academy of Social Sciences (VASS) to better inform policy; a review of Wellcome Trust's 'Livestock for Life' public engagement programme to improve animal welfare in the developing world. Although I am based in ODI's London office, my work has taken me to over 20 countries in Asia, Africa and Europe.

Ajoy Datta (UK), Zambia, Ghana, India, Vietnam, Indonesia

LAW AND DEVELOPMENT

The rule of law is of fundamental importance to development, promoting economic growth, democracy and human rights. However, strengthening the rule of law in developing and transition countries remains a major challenge, as in many countries multiple systems work side-by-side, each weakly enforced, and often operating in contradiction with each other. Creating a unified and robust system of law is one of the biggest challenges these countries face. Corruption can distort outcomes, and officials lack training. Favouritism is common and thus does not provide equitable protection to all citizens nor inspire their trust in the rule of law.

The work: You may be involved in empowering people in the developing world to access justice, economic opportunities and good governance. Your work could focus on the rule of law: ensuring legal systems work and that the law is enforced in accordance with its intention. You might also be working on international policy issues that relate to development, such as assisting and drafting legislation, redrafting or revising laws, or providing feedback in relation to new laws and development policy. There is also private international law or cross-border transactional law, where you might be advising on international treaties or on private transactions that cross international borders. Security and justice sector reform may also be a focus, or promoting investment climate reform and anti-corruption laws and policies.

The employers: Law reform in developing countries was initially sponsored by organisations such as USAID and the Ford Foundation. Today the range of employers for law and development positions is incredibly diverse. Lawyers without Borders or Avocats sans Frontières (different organisations) are a good place to start. Entities such as the International Court of Justice serve as the pre-eminent authority on disputes between states while a wide variety of NGOs address rule of law issues and assist with developing legislation and policies that align with development initiatives. Private sector employers such as corporate law firms and companies are often involved in complex cross-border transactions across the developing world that often have significant development implications for the host countries. In addition, in the public international law arena, you might work for one of the international criminal law tribunals, such as the International Criminal Tribunal for Yugoslavia.

Breaking in: Perfect your drafting skills – the biggest assets lawyers abroad typically have is not a developed knowledge of legal systems but the fact that they have excellent drafting skills and can easily translate and communicate the intentions of their clients in a clear and concise way.

Job titles: Legal Officer, Rule of Law Officer, Associate, Policy Advisor, Transactions Advisor, Legal Expert

Find out more:

Michael J. Trebilcock and Robert J. Daniels (2009) *Rule of Law Reform and Development: Charting the Fragile Path of Progress*, Cheltenham: Edward Elgar.

See also: Human rights; Protection; Governance

I studied Political Science with a double major in Japanese and went straight to law school after graduating. My main interests were transactional law and international law, both private and public. After law school I went to Vienna for a few months and worked for an international organisation focusing on European security matters. I worked in the legal affairs unit, focusing on diplomatic immunity, organisational status, and also more significant international issues such as security concerns and human trafficking. After this I then went back to New York to start at an international law firm. My interest was really to work internationally but I had student loan debts to pay off first, so wanted to focus on this for three years before taking a more international role. In order to ease the transition I involved myself in any type of work that could be relevant or offer transferable skills to an international environment. I got involved in cross-border-related transactions, and did a lot of work in project finance, which is a new model for development in a lot of countries. Many large new project financing companies have a PPP model, where part of the project will be funded by a large international organisation, such as a multilateral investment bank (e.g. ADB), and the other part by a commercial bank.

My firm had a connection with the ICTR so I immediately jumped at this opportunity and spent three months in Arusha, Tanzania. I was interviewing witnesses for an indictment; it was exciting to see how international law is practically applied beyond theories and treaties, and learn about command responsibility and other international criminal law principles. Being in the field I made contacts who were looking for attorneys with my background and thus I was offered work.

I now work for a government investment promotion institution, as a transactional legal officer, on a World Bank funded project. Part of my role is also working alongside a national counterpart to build capacity. My position specifically focuses on transactions that have been identified by the government as key areas that will contribute to the country's development (e.g. energy, ICT and mining). We negotiate and implement each transaction in line with the government's development initiatives and, most importantly, ensure that the final deal has a positive impact on the government's citizens.

For anyone working in the international law field, it is important to realise it is very challenging to get into. There are a few key organisations that many want to work with such as the UN, IFC, World Bank, etc. My advice is not to be too focused as this will narrow your options. Don't be too selective and aim towards anything that sounds interesting. You never know, it may open up a new world for you.

Jeannetta Craigwell-Graham (USA) Tanzania, Rwanda

LIVELIHOODS

How do poor people make a living? What can be done to reduce their vulnerability, improve their quality of life, meet their basic needs and escape poverty? To answer these questions you will first need to understand where their income comes from, how it can be diversified and what assets are at their disposal. In rural communities, livelihoods will be closely associated with food security, agricultural and livestock productivity. In other cases you may be working with vulnerable groups, such as widows, sexual and gender-based violence (SGBV) victims, young people, victims of gender injustice, to help their economic reintegration, engaging them in alternative income generating activities (IGAs) and access to microcredit. Supporting the development of people's livelihoods is of course a complex issue that involves several stakeholders, from government and policymakers, to investors and markets. The proverb 'Give a man a fish and you feed him for a day. Teach a man to fish and you feed him for a lifetime' applies very well to livelihoods work.

The work: The first step is always a situation analysis to understand the context: what are the determinants of poverty and what explains different poverty levels between locations and households. You might do an analysis of livelihood strategies (the different activities that people engage in to make an income) and map livelihood assets and opportunities. This will guide programme development and enable the design, coordination and implementation of sustainable livelihood activities that will reduce that person's and their family's vulnerability (to escape from poverty, famine, exploitation, etc.). You will frequently be involved in developing or supporting a livelihoods strategic plan, designing interventions – such as supporting the development of IGAs – coordinating community and regional responses, and be involved in monitoring and evaluating the impact, asking questions such as: have people become less vulnerable to food insecurity, poverty, exploitation? The work might also involve connecting people to markets, analysing value chains and identifying opportunities for income-generating activities, to conducting literacy projects and ensuring gender justice.

The employers: Many NGOs working directly in rural development, gender equality, food security and so on are involved with livelihoods. It is a very cross-cutting issue and is relevant both for development, post-conflict and post-emergency situations.

Breaking in: A Master's in Rural Development, Social Development, Agriculture, Economic Development, or Gender is a good start, going on to specialise in Livelihoods analysis for a Master's dissertation for example.

Job titles: Livelihoods Specialist, Livelihoods Advisor

Find out more: The Livelihoods resource centre of the IFRC is a good starting point. www.livelihoodscentre.org

See also: Food security. Microfinance; Private sector development

---

I studied Economic Sciences and Development at university in Bouake, Ivory Coast. I enjoyed the subject and after getting my degree I decided to do an MSc with specialisation in Management Sciences followed by a Diploma in Advanced Studies in Local Development.

After my studies I worked for some local organisations to identify and lead income generating activities for vulnerable populations. After one year with local organisations, I joined the INGO International Rescue Committee (IRC). My job focused on economic protection and reintegration of internally displaced people and the activities I led included supporting displaced people doing income generating activities and implementing cash for work interventions. From there I was promoted to manager of an economic recovery programme for access to financial resources where I set up 18 Village Savings and Loan Associations (VSLA) and provided guidance for income generating activities. A large part of this role also required me to build the capacity of local staff and be involved in monitoring and evaluation activities, ensuring that the project was implemented as planned.

With IRC I continued to grow professionally, being promoted to roles of greater responsibility, as National Manager for Economic Recovery and Development. In this role I was responsible for all economic recovery and development activities nation wide, and this involved developing IRC's national strategy, writing grants and proposals, providing technical advice to staff and monitoring and evaluating programmes.

After four years with IRC, I applied for my first international position with International Medical Corps (IMC) in the Democratic Republic of Congo. There I am responsible for the livelihoods component of a GBV response programme that provides medical, legal, psychosocial and livelihoods assistance to survivors of GBV.

Livelihood activities cannot just focus on income generating activities but need to also address the issue in a holistic manner. It is important for people who engage in livelihoods activities to have the necessary skills to earn an income, which include literacy, education and the ability to execute specific tasks, but it also means understanding the activities that are needed within the community. My job requires undertaking market analyses, and helping the women we work with identify what businesses are most needed. With IMC, I manage a livelihoods programme across four sites, and I need to support my staff at each site. I provide training, carry out field visits to supervise activities and provide feedback; I do market analysis to make sure that the activities the women are undertaking are viable. A large part of my work also consists of reporting, doing cash projections and ensuring that targets proposed to the donor are met.

I like my job very much. Although the context in DRC is challenging, I like supporting the staff and seeing how the women we work with become more independent financially. What I don't like is the insecurity, which makes it difficult to sustain some activities.

Livelihoods is an interesting area in development. It takes into consideration many aspects and supports people in leading better lives. Someone who is interested in livelihoods should choose his or her studies in the area of economic development or agribusiness – both are helpful to work in this field. I also recommend doing an advanced degree and specialising in a specific area.

Patrice Boa (Ivory Coast) DR Congo

LIVESTOCK AND VETERINARY CARE

Domestic animals play a vital role in the agricultural and rural economies of the developing world, and are essential to the livelihoods of millions of families. They are a direct source of food, support crop agriculture (providing animal power, fertilisers) and for many smallholder farmers livestock are the only ready source of cash available to them (to cover everything from crop inputs – seeds, fertilisers – to school fees, medicines). Small animals (e.g. ruminants, fowl and pigs) with high reproduction rates can provide a regular source of income, while larger animals such as equines, bovines and camelids are work animals that can help to plough fields, fetch water, carry wood, carry goods to the market and so on. Many animals are vital for the families' livelihood, or a capital reserve, to use in times of hardship or when facing large expenses such as a wedding or hospital bill.

The work: If you work in animal production and livestock you might be training farmers to improve breeding techniques or running a village livestock programme (e.g. one goat per family) or developing animal husbandry schemes. Another important area is disease control and surveillance in times of global epidemics. Animal welfare is also a key consideration in development: generally, the better the state of the animals the more productive they will be, and the greater impact and security their owners will have on their livelihood. Also, non-productive or working animals such as dogs and cats or wild animals such as bears are beneficiaries of health or neutering programmes. You could be supporting pastoralists to access markets and thus improve their capital or working with the ministry to support a fisheries project.

The employers: Many organisations working on livelihoods will also focus on livestock and animal production. Other organisations specifically addressing these issues include: World Organisation for Animal Health (OIE), FAO, Vétérinaires

Sans Frontières (VSF), Farm Africa, Send a Cow, Society for the Protection of Animals Abroad (SPANA), World Society for the Protection of Animals (WSPA), World Horse Welfare, The Donkey Sanctuary, Worldwide Veterinary Service, The Brooke.

Breaking in: For some positions a background in veterinary sciences is essential, while in others a general background in livestock, agriculture, animal welfare, agribusiness, agricultural economics is sufficient. International experience will be essential.

Job titles: Livestock Specialist; Veterinary Advisor; Food Security Specialist; Animal Welfare Specialist; Diseases Surveillance Expert

Find out more:

Jemimah Njuki and Pascal Sanginga (eds) (2013) *Women, Livestock Ownership and Markets: Bridging the Gender Gap in Eastern and Southern Africa*, London: Routledge.

See also: Livelihoods; Farming and agribusiness

I studied Veterinary Sciences in Catalunya and started working as a general vet after I graduated in 1997. In 2002 I moved to the UK and found that my veterinary training was very scientific and I wanted a broader perspective. So I did an online course in Humanities, specialising in animal welfare, a teaching diploma and now I am doing a diploma in International Cooperation).

It was a friend in the UK who sparked my interest in international development work – she worked for a health NGO and spent a lot of time travelling in Africa. I started to think that maybe I could use my skills in this area too. To improve my chances I did a Master's in my area of greatest interest, Animal Welfare and Behaviour, at Edinburgh University and also did three months of volunteering in Fiji, the Cook Islands and Samoa with charities linked to the World Society for the Protection of Animals (WSPA).

It was thanks to this experience – and a teaching diploma – that I secured my role with The Brooke, which is an animal welfare organisation dedicated to improving the lives of working horses, donkeys and mules in some of the world's poorest countries. Our philosophy is that if you improve animal welfare – happier and healthier animals – you also improve people's livelihoods.

I started as a Veterinary Training Advisor, developing a veterinary competencies framework training curriculum for the 150 Brooke vets scattered around the world, devising participative and student centred approaches and planning who trains where, when and how. Now I work as a mentor and advisor to The Brooke's partner NGOs: I support veterinary expertise, give technical support and advice on the programmes, and oversee animal welfare and how they do what they do and try to do. Our partner NGOs are based in South and Central America, Africa and Asia.

I am based in London but go on eight or nine trips per year. My most recent trip was to Guatemala. Here, equines are an important source of the family economy, but this contribution to livelihoods is not often recognised and equines have a rough deal of it. In many remote communities there are no vets – a vet's visit sometimes costs about the same as the animal itself so they never call one. Here we work to train community members in basic animal health, husbandry and the soft approach to animal handling. During visits I support our local service provider ESAP (Equinos Sanos para el Pueblo) in areas ranging from veterinary expertise to programmatic issues. Part of the visits involve training in health issues and also trying to find the most effective and efficient way to improve welfare (through education, specific interventions etc.). In Peten for example there is a big problem with ticks. But in Chimaltenango they complained that the animals, especially the mares, are very thin after giving birth. By observing, I noticed that they were giving them different foods, which they thought were more nutritious but in fact are low in calories, which could be a reason. Sometimes, as an outsider you can see what the people living in the situation cannot see, but I definitely learn during each visit at least as much as people learn from me.

Josep Subirana (Spain), Senegal, Egypt, India,
Nicaragua, Pakistan, Guatemala

LOGISTICS

Logisticians oversee the management of resources from the point of origin to their destination. These may include emergency food aid, materials for shelter, high-end medical equipment or drugs to equip a hospital. Logistics is a core component of humanitarian support, getting the required aid out to those who need it the most as quickly as possible, and accounts for a very large proportion of the costs involved in disaster relief. The contribution of logistics to the delivery of aid – getting the right goods, to the right place, at the right time – quite simply saves lives. In a development context there often isn't the same level of urgency but the role is equally important – sourcing the resources and equipment necessary to implement a project. A core component of the role involves understanding the supply chain and exploring ways of minimising the costs while maximising the impact of the aid.

The work: The logistician will be directly involved in planning warehouses, controlling inventories, procurement (which often includes dealing with port operations), supply chain, field finance and audit trail. In an emergency context, the logistician plays a key role in planning and managing the response. Logisticians will be involved in disaster preparedness and will feed into emergency plans before a disaster actually occurs. They will look at how to maximise the response and minimise the distribution time, money and number of distribution centres required. Special software and mathematical models are often used to do so.

The employers: Logisticians are in demand in both humanitarian organisations that provide support during an emergency response and large-scale development projects focusing on issues such as health or security. The employers for the former are likely to be NGOs or multilateral organisations such as UNHCR. In the latter you could be working for a development consultancy, government, INGO or other multilateral organisation.

Breaking in: You can start with a logistics qualification or train on the job in either the private sector or by targeting a development project with a large logistics component. To move across to humanitarian logistics you could gain exposure to the rigours of field and headquarter logistics through volunteering. To work in the field you will need training – several Master's programmes exist in Logistics and Humanitarian Emergencies – and prior experience of humanitarian work. Some organisations run training programmes: Save the Children UK run a one-year Logistics Skills Development Programme (LSDP). The Certification for Humanitarian Logistics (CHL) is a course designed/developed by seasoned practitioners in the sector for the sector.

Job titles: Logistics Coordinator; Warehouse Manager; Supply Chain Specialist; Procurement Officer

Further information: www.humanitarianlogistics.org is a website specifically for those working in the field, or wanting to train in the field and highlights conferences, events and training resources for those working in development. Humanitarian Logistics – a career for women? www.wise.uk.net/publications/Humanitarian_ Logistics.pdf

See also: Procurement

Growing up in Western Kenya, with a single mother who had to struggle to support her young family, gave me first-hand experience of the challenges of living in a developing country.

I believe in lifelong learning and while working in Nairobi, I earned the First Certificate in Purchasing and Stores in 1986 and in 1998 I successfully completed the Foundation Stage of the Graduate Diploma admitting me as a member of the Chartered Institute of Purchasing and Supply (CIPS), UK. Subsequently I studied for an MBA, graduating in 2003. By 2004 I was elected a Chartered Member of the Institute of Logistics and Transport (CILT). Currently I am a doctoral student researching public health supply chains in developing countries to identify lessons to improve health outcomes.

My first jobs were in the commercial sector but I always had my eye on humanitarian work. Then in 1994 I joined World Vision's Sudan and Somalia Programme as Procurement and Logistics Manager and later the ICRC. These jobs gave me a sense of purpose and I decided to devote my career to humanitarian work. In 2000 I became operations Manager for Oxfam GB's Regional Office in Nairobi. Our field staff worked in a high risk environment in South Sudan where they regularly came under bombardment which I also experienced first-hand.

In 2002, I was offered the position of Global Logistics Advisor in the Humanitarian Department at the Oxfam GB headquarters in Oxford. It meant moving to the UK with my two daughters to start life in a new country. I was very fortunate to be one of the people selected to benefit from mentoring from an Oxfam consultant who had developed the organisation's logistics and supply strategy. This provided me with invaluable management tools and a deep understanding of professional competence. It was a turning point in my career and enabled me to rise to Acting Global Head of Logistics and Supply. In that role, I led the Oxfam logistics team during two of the biggest humanitarian crises of our time: the Darfur crisis and the Indian Ocean tsunami.

Looking for new challenges I subsequently joined UNFPA, and later UNICEF, in Copenhagen, helping them to optimise their logistics and supply chain systems for development and emergency response programmes. In 2012 I ventured out on my own as an independent humanitarian and development supply chain consultant.

Logistics is a male-dominated field but I encourage women to consider this rewarding career. A day in the life of a field logistician can be challenging during an emergency. The logistician in the field is faced with managing the supply of goods in a complex, chaotic, and sometimes dangerous, situation. Headquarters staff support the field operations. When I was responsible for global humanitarian logistics operations, I would be in daily contact with field staff during an emergency operation, addressing their needs, providing leadership and problem solving to ensure that the operation ran smoothly. Making quick decisions is a must as there is little time to reflect during a crisis.

Pamela Steele (Kenya) South Sudan, Sierra Leone,
DRC, Pakistan and North Sudan

MATERNAL, CHILD AND REPRODUCTIVE HEALTH

Women and young children still carry the heaviest burdens of death, disease and disability in the world. Millennium Development Goal 4 was to reduce the under-five mortality rate by two-thirds (between 1990 and 2015). Some important steps have been made in this area. For example in 2011, 6.9 million children under five died, compared with 12 million in 1990.[1] This has been achieved in part through roll-out of immunisation campaigns including measles (which accounted for about one-fifth of all deaths) and the uptake of simple hygienic practices, such as hand washing and basic essential new-born care techniques. Millennium Development Goal 5 aimed at reducing by three-quarters the maternal mortality ratio and also achieving universal access to reproductive health. Ensuring that births are attended by skilled personnel is crucial for reducing perinatal, neonatal and maternal deaths. This has been achieved in some regions but in Africa this figure remains at less than 50 per cent. But an estimated 222 million women with an unmet need for family planning contributes to pregnancy and childbirth remaining the number one killer of girls 15–19 years old globally. Enabling women to decide whether they wish to have children, when and how many is still a long way off for many.

The work: You might be overseeing child health programme design, implementation and quality assurance; or working on the roll out of a vaccination campaign. You could also be ensuring that the quality of child health programme implementation is consistent with WHO recommendations. As an advisor you might be supporting informed decision making around organisational or national health programmes, guiding strategy development, supporting institutional learning and engaging in international discussions around reproductive, maternal, neonatal and child health.

The employers: From the WHO and UN, to bilateral advisors, INGOs such as Planned Parenthood Federation, Save the Children and Marie Stopes International and local Civil Society Organisations. Donors such as the Bill and Melinda Gates Foundation, government departments for international development and ministries of health, women and children also have a significant emphasis on reproductive, maternal, neonatal and child health.

Breaking in: On the programme side you may not necessarily need to have a background in the field; for advisor and senior level jobs a post-graduate qualification or medical degree may be necessary in addition to several years' experience.

Job titles: Maternal Health Specialist; Reproductive Health Advisor

Find out more:

Cathleen Miller (2013) *Champion of Choice: The Life and Legacy of Women's Advocate Nafis Sadik*, University of Nebraska Press.
United Nations (2009) Tracking Progress on Child and Maternal Nutrition: A Survival and Development Priority Within Our Reach.

See also: Public health; Health professional; HIV/AIDS, tuberculosis and malaria

---

If I look back, my career looks as though it has followed quite a clear path. But it wasn't until I finished university (a degree in Sociology) that I found the area I wanted to work in. Even then the focus was still broad, based around a desire to help others, a strong urge to be part of different cultures, working in different countries, and a sense of the devastating impact HIV was having globally. After travelling and doing temporary work with charities I secured a place on an MSc in Reproductive and Sexual Health Research at the London School of Hygiene and Tropical Medicine – a phenomenal course.

This was followed by an eight-month internship on a UNAIDS project in London. Then, based on several people's advice, I went to volunteer for an NGO in Ghana's Volta Region, working on a qualitative research project looking at the causes and consequences of teenage pregnancy among school children. Five months turned into three and a half years. We gained funding from UNFPA and I was given the opportunity to set up and run the organisation's sexual and reproductive health programmes: community-based awareness programmes around teenage pregnancy and reproductive health, research on the impact of teenage pregnancy for young women and a youth centre offering information and support around sexual and reproductive health issues.

I am now based in London and work as a Health Advisor for BBC Media Action. I support organisational learning, strategy and business development, as well as project design, implementation and evaluation for teams in Africa and Asia across a variety of sexual, reproductive, maternal, neonatal and child health issues and HIV. I love the variety of my work. Most of my time is spent in London but with some travel:

one week I may be editing reports, doing a needs analysis on reproductive health issues; the next designing training for production teams in South Sudan or writing a proposal in Nepal. So far I have travelled and supported work in South Sudan, India, Bangladesh, Ethiopia, Nepal and Uganda.

My advice is to follow your heart, but at the same time be sensible: consider the jobs available, look at the job descriptions and work out how you can build the skills you need to get that job. Experience is invaluable as the field is incredibly competitive. Internships have become the norm, but if you cannot afford to be an intern consider volunteering after work or doing a couple of part-time jobs to try different things out (I had three a few years ago – a bit much, but it taught me a lot!). If you don't have a strong sense of what you want to do, try a few things out and see what feels right. Keep in touch with people you meet who have similar interests – you'll find that within your sector it's a small world and you never know when you might cross paths in the future.

Genevieve Hutchinson (UK) Ghana, South Sudan,
Ethiopia, Bangladesh, Nepal

Note

1   www.who.int/mediacentre/factsheets/fs290/en/ (accessed 30 January 2015).

MEDIA AND JOURNALISM

Journalists play an important role in creating public awareness of development, humanitarian and human rights issues. Editors and journalists serve as the gateway between professionals working in the field, the people they are trying to help (or should be trying to help) and the general public (donors who also have an increasing role in advocacy and campaigning). Communications for development are important both to raise awareness of issues and mobilise the public to put pressure on their governments, and to influence donors' priorities and where they spend their money, enabling the end beneficiaries to speak up about their concerns and what is important to them, supporting a bottom-up approach.

The work: You could be working in film, radio or print. You will be identifying relevant and timely themes, researching and developing the story. If you are working for a implementing agency (NGO for example) you could be documenting and reporting on their work and its impact. If working for a newspaper, magazine or website, you might be identifying and researching untold and relevant stories. Typically, you will be focusing on the human side of the story, while bringing in wider policy issues and implications. A good journalist must not only describe, but also delve deeper into issues, and explore how big policies are affecting developing countries and people on the ground. International development is complex and uncertain so a good journalist must appreciate and explore the interplay between fields such as education, environment, health, governance, culture and economics.

The employers: Organisations, especially NGOs, are increasingly employing journalists, photographers and film makers (either as staff or freelance contractors) to convey clear messages to their donors and the wider audience. A number of print and online publications and websites report exclusively on issues of poverty and inequality.

Breaking in: An interest in current affairs and being able to identify tomorrow's top story is essential, as is knowledge of development, humanitarian and human rights issues. An interest in investigative journalism and getting to the bottom of issues is vital. You don't necessarily need to have a background in journalism or film/documentary making, although it does help if you want a full-time career in this area. Many publications welcome freelance contributions from people already working in the sector so this can be a good way to start. You can also volunteer for an organisation, writing or making a film about their work. *The Guardian* (UK) runs a yearly international development competition for unpublished and published journalists. Deadlines are usually in May.

Job titles: Journalist; Editor; Freelance Journalist; Contributor; Writer

Find out more:

Herman Wasserman (ed.) (2010) *Popular Media, Democracy and Development in Africa*, New York: Routledge.
www.newint.org
www.panos.co.uk

See also: Communications; Advocacy and campaigning; ICT

In my early twenties I visited Nigeria and the questions came thick and fast. How come a country that was so rich in oil wealth still had so many desperately poor people? Why was a country that was proudly independent, dominated by Western multinationals? Why was corruption so rife and what could be done about it? The answers I got were evasive and unsatisfactory. But I knew these were important questions to ask. I also knew, at last, what I wanted to be. I returned to England with that goal in mind.

I started working for a local newspaper, received training and freelanced. Three years later I'd saved enough money to realise my dream – to go to Latin America and try to work as a journalist there. After two months travelling I arrived in Peru and showed up at the *Lima Times*. The editor said: 'Okay, there's a typewriter. Write us a series of articles based on your experiences so far.' I did so, they were published, and I was given a job as a roving feature writer. For two years I travelled across the country, covering stories such as the internal war between Shining Path insurgents and government forces, the cocaine trade, mining in the Amazon and the plight of indigenous people. Underlying all these stories were the issues of equality and development – or lack of them. One day I came across an issue of the *New Internationalist*. I thought: 'This is the magazine I've been looking for! This tells it as it is.' Little did I know then that it would be the magazine that I would end up working on, as a co-editor, for many years to come.

New Internationalist covers stories that are not widely reported in the media. We often source stories from the global South and commission writers and photographers from there. We are a non-hierarchical, workers' co-operative. Each month the editorship of the magazine rotates a different co-editor. Once every two or three years I will get the chance to investigate, write and take photographs for the main theme, usually travelling abroad to do so. I also get involved in writing or editing *New Internationalist* books (for our No-Nonsense Guide series, for example), and working on contracts for NGOs and UN agencies. Part of my day may be taken up with commissioning journalists or illustrators, editing, writing and researching. I also need to keep abreast of social justice and development issues and helping to run the co-op takes quite a bit of my time.

If development journalism is your goal, initiative and commitment are more likely to get you noticed than long lists of academic qualifications. Knock on doors and show your face – you may get the brush-off but you never know . . . Then get some experience working in the Majority World. Write when you can and as well as you can. With the Internet, it's so much easier to be published today, but keep an eye on the quality. Sloppy writing may stay around, up there – with your name on it – forever. Don't let it define you.

Vanessa Baird (UK), particular interest in Latin America

MICROFINANCE

Microfinance is the provision of financial services (loans, savings, money transfer services and insurance) to the poor. This group is typically excluded from traditional banking facilities, where they are not considered as a viable market. The poor are nevertheless economically very active, and access to financial services offers them opportunities to improve their lives, empowers them to build and expand small businesses, lifting them out of poverty and reducing their vulnerability. Modern-day microfinance was pioneered by the Grameen Bank in 1976, which began lending to poor women in Bangladesh. Microfinance institutions (MFIs) blossomed in the 1990s. Microfinance, especially microcredit, has been elevated to celebrity status,[1] with Grameen's founder Muhammad Yunus being awarded the Nobel Peace Prize in 2006. Microcredit really does seem to work for some, as repayment rates tend to be high, although it is not a *silver bullet* for lifting people out of poverty.

The work: Working for an MFI you might be developing and testing financial products for your clients, monitoring and overseeing repayment rates and rolling out services to marginalised communities. You could also be forming partnerships with commercial co-financiers or guarantors, performing market soundings and providing risk appetite feedback. It is essentially banking aimed at the poor, and you will need to understand the needs and concerns of your clients. On the ground you could be supporting and training groups on how to access and manage microcredit (often done by local staff due to language barriers) and finding local partners who can manage it. Working for an NGO you might be helping to set up VSLAs or helping train people to be ready on how to access microfinance and build their businesses. The preference is usually to administer finance in a group setting, as repayment rates of groups tend to be higher than for individuals.

The employers: As a development practice, microfinance appeals both to the right (entrepreneurship and credit) and the left sides (empowering women) of the political spectrum. This is also reflected in the range of MFIs you could work for – from small non-profit organisations to large commercial banks. Several other such MFIs now exist and the Kiva website www.kiva.org lists their field partners, from big to small MFIs.

Breaking in: A background in finance and banking or a numerical-based discipline, and/or a general interest in this area (focus your Master's dissertation on microfinance) will put you at an advantage. Many MFIs offer internships in this area, among others the Kiva Fellows scheme. www.kiva.org

Job titles: Access to Finance Advisor; Microfinance Manager; Repayment Officer; Loan Analyst

Find out more: http://cgap.org CGAP is an independent policy and research centre dedicated to advancing financial access for the world's poor, housed in the World Bank.

Ananya Roy (2014) *Poverty Capital: Microfinance and the Making of Development*, New York: Routledge.
Ian Smillie (2009) *Freedom From Want: The Remarkable Success Story of BRAC*, West Hartford, CT: Kumarian Press.

See also: Economist; Private sector development; Livelihoods

After eight years of doing management consultancy work in England it was time for a change. I wanted to work in development and luckily my employer in the UK was supporting a microfinance project in Kenya that I got the chance to help set up. That sparked my interest in microfinance. To facilitate my transition into the field, I completed a Master's in Rural Development at Sussex University and this project became the subject of my dissertation. I got the chance to interview a senior person within an MFI in Kenya. A year later I wrote to him to see whether there were any employment opportunities for me with the bank, and that is how I first ended up on a six-month contract working as Special Projects Officer with Opportunity Kenya. Six months became two years.

My background in consulting set me up well for the job, except this time I am not working for external clients but on internal processes. This is more satisfying as you can actually see the fruits of your work and oversee implementation. Some of the most useful competencies that I brought with me were project management skills, planning, delivering clear presentations, data analysis and financial models. I am currently Associate Director of Operations, Urwego Opportunity Bank, working on extending outreach and financial services for the poor through making mobile banking more convenient and secure. While microfinance is profitable, it is not as profitable as conventional banks and does not cover all our costs. We need to seek funding from donors and consequently one of my responsibilities is also writing proposals and reports for donor funding, so that we can extend our services to areas that are not so profitable and make our services accessible to the poorest.

I had this vision that 'working in development' would be exotic. But essentially the day-to-day work is similar to my life back in the UK: I spend much of my day in the office working on my laptop or in meetings. But I do find the work much more satisfying and I try to spend about a quarter of my time out in the field as I need to understand our clients and the situation for staff working out at the branches. This gives me the opportunity to meet the people we are actually helping, to see what is working. The challenge though is to really see, and measure, the impact our work has on their lives. Microfinance too often just focuses on the numbers, how people's finances have improved, and not how their life has improved. But I still find it a very satisfying and rewarding job and appreciate the chance to work for an organisation whose mission, vision and values I truly believe in.

Nicholas Meakin (UK), Kenya, Rwanda

Note

1   U2 lead singer and celebrity humanitarian Bono is reported to have said 'Give a man a fish, he'll eat for a day. Give a woman microcredit, she, her husband, her children and her extended family will eat for a lifetime.'

MIGRATION

Migration and displacement are pressing issues during conflicts and disasters, when large populations are forced to flee from their home environment in search of security. But even poverty and lack of employment opportunities push people to take risks, travelling far from their home looking for a brighter future. These people are vulnerable and thus easy targets for trafficking and forced recruitment. Women and children are particularly vulnerable, being pushed into prostitution, drug trafficking and domestic work, but men are also victims, tricked into doing unpaid or low paid work. This can have serious implications on human rights. On the flip side, migration can also have a very positive role on development, through diaspora remittances and skills transfer. Official remittance flows to Africa became the largest external financial source reaching a record $60.4 billion in 2012 – overtaking foreign direct investment and official development assistance (African Development Bank).

The work: Organisations working in this area may focus on regulating and facilitating migration, addressing forced migration and promoting resettlement and reintegration. The field is very cross-cutting and may require expertise in international migration law, policy analysis and advocacy, protection of migrants' rights, migration health and the gender dimension of migration. The work can range from hands-on work in an emergency context to policy level. You may be managing a project on human trafficking, working to provide people vulnerable to exploitation with livelihoods to reduce their vulnerability, or reintegrating returnees both socially and economically back into their communities. At the other end of the spectrum you may be helping to prevent 'brain drain' or engaging diaspora to contribute to economic or service development back in their country of origin.

The employers: The International Organization for Migration (IOM); UN High Commission for Refugees (UNHCR); United Nations Office against Drugs and

Crime (UNODC), also working in human trafficking issues; International Labour Organization (ILO); International Centre for Migration and Health; International Network on Displacement and Resettlement; Migrants' Rights Network; diaspora groups working in development; Migration for Development in Africa (MIDA) are just a few.

Breaking in: A course or Master's in Refugee and Migration studies, will give you a background. There are many courses such as SOAS (UK) MA in Migration and Diaspora Studies; Oxford University's (UK) MSc in Refugee Studies and Forced Migration; the University of the Witwatersrand (South Africa) has an MA in Forced Migration Studies; York University (Canada) has a Refugee and Migration Studies Graduate Diploma. The American University of Cairo has a Center for Migration and Refugee Studies; and Georgetown University (USA) has the Institute for the Study of International Migration. After you are armed with an academic background, it is mostly learning by doing. Try to get experience with an organisation that works in this area.

Job titles: Resettlement Manager; Diaspora Engagement Officer

Find out more:

www.refugeeeducation.com/

See also: Refugee camp management; Livelihoods

Growing up in Uganda I have always been interested in women and development. I did my undergraduate degree in Canada in Political Sciences and Women's Studies: the feminist methodologies came as a surprise to me as I was more interested in gender and development, but provided an interesting perspective. I later did a Master's in Women's Studies with a Graduate Diploma in Refugees and Migration Studies. That is what I was really interested in and when I returned home to Kampala I was lucky to get an internship at IOM through a contact, and five years later I am still with them.

I have now worked with IOM in Uganda, Kenya, Rwanda and the Maldives. I work both with migrant workers, irregular migration (human and child trafficking) and emergency post-crisis migration. In Kenya, for example, I worked with pastoralists whose livelihoods had been altered by climate change, so they were vulnerable to exploitation and trafficking. The work can involve direct assistance, looking at the root causes of the problem, raising awareness of the issues among partners and governments, engaging partners, counselling, skills development and reintegration; essentially, the full cycle of services to get the victims of human trafficking reintegrated back into their communities, with alternative livelihoods to reduce their vulnerability. I enjoy it because of the direct involvement with local communities and individual people, but also the higher-level work, such as policy dialogue with the governments, training partners, lobbying government and motivating other organisations to share the concerns and get involved.

In Rwanda I helped set up a counter-trafficking project – the first of its kind. We signed a partnership with the government, raised awareness and trained stakeholders. We used the media, particularly radio and TV adverts, to advocate safe migration and the dangers of irregular migration. Currently, I am based in Maldives where I am in charge of establishing IOM's presence in the country. The Government of the Maldives (GoM) specifically requested IOM to set up an office here, assisting the government with migration management, particularly looking at human trafficking. Maldives is unique country in the region as it is a destination country for unskilled migrant workers. Many face forced labour conditions with long working hours, non-payment of wages and withholding of travel documents. Some women are also subjected to sex trafficking while children are trafficked to the capital Male from other islands for domestic work. We are providing the GoM with technical assistance in all key areas of intervention. We are working in all areas including prevention, prosecution, protection and partnerships.

What I enjoy most about the work is seeing the direct results: working with people who have nothing in a desperate state, crossing a border with pots or pans, walking barefoot across the hot sand, and actually being on the ground to be able to help them and get them back to a stable life. They appreciate all the help we can give them.

Alia Hirji, (Canada/Uganda), Uganda, Rwanda, DRC, Maldives

MONITORING AND EVALUATION (M&E)

M&E has become a buzzword in the sector: everyone is talking about it, doing it and even evaluating the evaluation. Widespread lack of accountability is commonly seen as one of the failings of development efforts so this relatively new emphasis on M&E is one way to enhance transparency, impact and cost-efficiency. What outcomes is the project having? What changes can we observe on the baseline data? Is it achieving what it set out to achieve? If not, why and what can we learn from this? While often referred to together, monitoring and evaluation are synergistic but quite distinct activities. Monitoring is the ongoing, systematic collection and use of data about a project to inform ongoing management and improvements to the work. Evaluation assesses the worth of the project through periodic reviews and provides the means for learning from past experience, so as to improve programmes and demonstrate results to key stakeholders. The challenge is that organisations are under pressure only to publish successful results, as negative outcomes might jeopardise future funding, but this is often where the learning lies.

The work: M&E starts right from the project development phase: planning how to measure the project's baseline (the status at the start of the project), outputs and outcomes (what the programme is trying to achieve) and indicators (what will be measured) through the systematic collection and use of data. It might also mean developing the tool for data input and analysis. Evaluation involves the analysis and interpretation of the data, and the recommendation to improve the programme. If you are working in-house you may be involved in all elements. As an external consultant doing the final evaluation, you will be brought in just to do this. Different types of evaluations exist: for example, impact evaluation, process evaluation (a randomised controlled trial), cost–benefit and cost-effectiveness analysis.

The employers: Agencies implementing projects but also donors who want to understand the impact of their funding. There is an increasing expectation of understanding these methods for all development workers, whether it is a core component of your role or not. In recent years there have been significant attempts to institutionalise M&E and now it is at the core of most organisations, which is reflected in the number of jobs in this sector. There are in-house jobs, as well as consultancy jobs for external evaluations and M&E advisors. Many donors also have designated people to support the organisations they are funding to improve their M&E systems and improve accountability.

Breaking in: One route in to M&E is to do a Master's course with a strong emphasis on both qualitative and quantitative research methods (in a social science, economics or health for example) and do a research-based dissertation out in the field. Get contacts and you might be able to land a social research or M&E internship or job after you graduate. A mathematical or statistical background will help you to specialise in the quantitative analysis, a good grasp of IT will also put you in a good position to develop effective monitoring systems. Social scientists and anthropologist are also good at the more qualitative aspects. M&E is a core aspect of many jobs, enabling you to learn on the job and direct your career in this direction. Some specialist jobs may require a PhD.

Job titles: Monitoring and Evaluation Officer/Manager/Consultant

Find out more:

Making Monitoring and Evaluation Systems that Work: A Resource for Development Practitioners, World Bank Publications (Feb 2010) ISBN-10: 0821381865

See also: Programme management; ICT

My parents had travelled in Africa so when I was young I always had an interest in the area. At university I specialised in African History and before graduating I sent out several prospective applications to development organisations. Only one responded, a friend's dad, who was a development consultant. I worked with him for a year and this is where I first learned about LogFrames.[1]

From here I went on to CAFOD, based in London, where my job was primarily grants and proposal writing. My LogFrame experience was really useful here, as most proposals require one. One aspect of these is *indicators*, and how we plan to measure the extent the project is achieving its aims. This all ties in to M&E. This is also where I learned about information systems and how to get the information donors needed to assess project proposals. I joined VSO as a volunteer in Zimbabwe and then returned to England to do a Master's in Social Anthropology of Development. This is relevant to evaluation as the course includes a large component of social research methods, which are used to some extent in evaluation. After my Master's I moved on to Concern Worldwide, was based in Uganda and my role was about

50 per cent M&E. From there I went to the International Rescue Committee (IRC) as M&E Manager based in South Sudan where I did a lot of field work supervising data collection. After this role I went on to become Quality Improvement Director for a health project in Tanzania, where my role is more office based, building systems and training up other staff.

In my field-based position in South Sudan one of my roles was to collect baseline information at the start of the programmes so as to be able to measure any changes the project would bring about after a period of time. For example, for one baseline survey, I trained up a team of local interviewers to collect data using interview guides. I often went out to the field with them and helped to address some of the problems they were encountering. For example, as there are no house numbers or street maps in many areas, it is difficult to do random sampling. So we would select a route through a village and interview people in every seventh house. I collected up the surveys each day and checked them for any inconsistencies or unclear information. After the month of data collection the forms were sent to Nairobi for data entry (since there was not a reliable-enough electricity supply or sufficient computers in the field) and a consultant was hired to do the analysis and write up the findings.

What I like about the work is getting into the field, sitting in people's houses as the interviews are being conducted, meeting people. I also enjoy training and mentoring staff, and seeing them develop and grow into a role. Handling expectation is challenging though. It is unrealistic to expect this type of evaluation to have the rigours of academic work. We are limited by time, resources and budget, so it's difficult to conclusively demonstrate impact.

Mya Gordon (UK), Tanzania, Zimbabwe, South Sudan, Uganda

Note

1   Logical frameworks: a matrix for illustrating project aims, objectives, outputs, outcomes and activities commonly used in project management.

NATURAL RESOURCE MANAGEMENT (NRM)

Many of the world's poorest countries are rich in natural resources – minerals, land, water, fisheries and forests – which need to be well managed and protected. If these natural resources are governed and harnessed in the right way it could be transformational, allowing governments to convert more of their natural resources wealth into inclusive and sustainable growth and development while at the same time protecting and conserving their ecosystems. Sadly though, investments in the timber, oil and mineral sectors have rarely translated to lower levels of poverty in the poorest countries. Overexploitation and unsustainable extraction methods along with population pressure have resulted in rapidly degrading land, forest and water resources, loss of biodiversity and mineral wealth, inadequate access to energy and progressive climate change all over the world, with people everywhere suffering from this destruction of natural wealth directly or indirectly.

The work: Addressing these challenges means strengthening local/global institutional arrangements for conservation and use of natural resources. From a *legal perspective* you might be advising on how to secure better terms from contracts in extractive industries, how much to extract, how to mitigate corruption and over exploitation and engage in informing policy. From an *environmental perspective* you may be evaluating the impact of natural resource use and extraction, and finding solutions to environmental and social risks. This might include work as diverse as the implementation of REDD+ in forestry projects, conservation of fishing grounds, field-testing of technological innovations in the renewable energy sector – or even conservation/environmental education. From an *economic perspective* you may be developing upstream and downstream value chains for extractive industries or assisting countries to develop optimal fiscal instruments and revenue management mechanisms.

The employers: Natural resource management professionals are recruited by:

1) bilateral/multilateral organisations and donor agencies such as United Nations Environment Programme (UNEP), USAID, European Commission, FAO, etc.; 2) international organisations/think tanks such as IUCN, International Water Management Institute (IWMI), Conservation International, Institute for Global Environmental Studies (IGES), World Resources Institute (WRI), etc.; 3) national government agriculture/irrigation/environment departments and extension services; and 4) national/regional NGOs, CBOs, universities, research institutions and other civil society advocacy organisations.

Breaking in: Specialising in the natural resource management sector generally requires a post-graduate degree in Environmental Science, Natural Resource Management, Forest Engineering, Energy Management, Geospatial Sciences, or Environmental Economics, etc.

Job titles: Natural Resources Programme Officer; Economic Advisor on Natural Resources; Land/Water/Resource Manager; Fisheries/Forestry Manager; Geographic Information System (GIS)/Remote Sensing Analyst.

Further info: Websites of international organisations such as UNEP, UNDP, FAO, IGES, IWMI, International Union for Conservation of Nature (IUCN) will help in finding out more about this sector.

See also: Livelihoods

I have been working in the field of environment and development for around ten years – mostly in a developing country context (India) with some industrialised economy exposure. I chose to position myself at the interface of the environment–development sector so that I could bring to each sector the perspective and experience of the other. I specialised in NRM by orienting almost all of my coursework, internships and work experiences towards that goal. My Bachelor's degree was in Geography and my Master's degree was in Environmental Science and Policy.

I started out as a Research Associate with Centre for Environment and Development, Calcutta. Here I participated in socio-economic surveys and environmental impact assessments (EIA) of infrastructure development projects. We worked with tea estates, wetlands, canals, river-beds and forest areas. I then joined Catholic Relief Services (CRS) Eastern India, as Programme Executive – Agriculture. My work involved participatory planning, monitoring and evaluation of rural micro-watershed projects, including resource mapping and watershed delineation; peer review and appraisal of several other NRM/Emergency Relief Projects. Most of the work was in rural dryland areas with marginalised populations that CRS worked with.

I took a mid-career break as a scholar-practitioner at Clark University, USA in their programme in Environmental Science and Policy. Back in India, I worked on short stints in environmental education, documentation and planning with a few organisations before joining Centre for People's Forestry (CPF), where I worked on the design and coordination of research initiatives in issues such as forest-based livelihoods, forest resource assessment, timber, fuelwood, fodder-based enterprises. In 2009 I started working as an independent researcher and consultant with several organisations such as International Water Management Institute (IWMI), Access Livelihoods Consulting (ALC) and the University of Hampi.

When working on NRM projects, I travel to project locations, interact with various stakeholders, and carry out focus group discussions, action research and capacity building activities. The desk work includes analysis, interpretation and integration of various components into comprehensive reports, proposals and studies. Carrying out an NRM project also includes walking or travelling long distances for ground surveys, mapping and preparation of topographic transects, delineation of project areas, assessment of physical aspects of project areas, resource appraisal etc. When working on GIS/remote sensing projects, I have to spend time poring over maps, topo-sheets, satellite images etc., and work with GIS/remote sensing (RS) software for long hours.

You need to have enthusiasm for the outdoors and a tremendous interest in how people and ecosystems interact with each other. Without that focus, it would be difficult to carry on, year after year, if you miss the comforts of home on every field trip and are unwilling to delve deeper into environment/development issues that probably don't concern you directly. You should also be comfortable with the knowledge that you will probably never be a millionaire working in this sector.

Job vacancies usually don't show up in newspaper vacancy sections, and you have to put in quite a bit of effort to get work assignments in your chosen sphere.

Shreela Chakrabarti (India)

NUTRITION

Malnutrition is one of the top five killers of children under five. Current research suggests that every year 13 million infants are born having suffered growth retardation in the womb due to inadequate maternal nutrition. Fifty-five million children under five suffer from wasting, 178 million from stunting, and there are 3.5 million unnecessary maternal and child deaths every year directly attributable to under-nutrition.[1] Public health nutrition is an important area of work in developing countries even when a quality diet may be available as malnutrition is far more complex than non-availability of food, and effective treatment requires a multi-sectoral approach. Malnutrition is exacerbated during humanitarian emergencies, especially when there is an accompanying disease outbreak and/or food insecurity.

The work: Following complex emergencies, the work might involve assessing and managing current nutritional problems. Surveillance of the nutritional situation according to anthropometric indices, as well as adequacy of household level food consumption could also be key components of the work. You might be developing a nutrition response strategy, liaising closely with health and food security experts or delivering community level nutrition training. Where a quality diet is available, response is based on integrating nutrition counselling related to optimal infant and young child feeding and improving quality of complementary foods through public education campaigns and behaviour change. Conditional cash transfers are one method that is used for participation in nutritional education.

The employers: Many humanitarian agencies will employ nutrition advisors as will development organisations working on food security and with children. Some examples include GOAL, UNICEF, Save the Children, etc.

Breaking in: A dietician degree is rarely needed for public health response; most often requirements include a MSc in Nutrition or a Master's in Public Health (MPH) with a nutrition specialisation or BSc with nutrition focus. You will often need some field-based experience, with work in a humanitarian organisation an advantage. The London School of Hygiene and Tropical Medicine for example has a Public Health Nutrition Master's programme.

Job titles: Nutritionist; Nutrition Specialist; Nutrition Advisor; Infant and Young Child Feeding Specialist; Social and Behaviour Change – Nutrition/Health and Nutrition

Find out more: www.nutritionworks.org.uk offers training, and holds a web-based register for International Nutritionists and Food Security Specialists.

UNICEF run the nutrition cluster www.unicef.org/nutritioncluster/ and several training courses. They also run training www.unicef.org/nutrition/training/

See also http://fantaproject.org/ and the Enough Food If campaign http://enoughfoodif.org/

The University of Southampton offers a free ten-hour online course primarily aimed at health practitioners focusing on caring for infants and children with acute malnutrition www.som.soton.ac.uk/learn/test/nutrition/Default.asp

See also: Food security; Public health; Advocacy and campaigning

---

I worked as a Peace Corps Volunteer in Nepal and became interested in the need to improve the quality of the diet of the students I was teaching. Wanting to specialise, I opted to study an MSc in Food Policy and Applied Nutrition at the Friedman School of Nutrition Science and Policy at Tufts University in Boston, Massachusetts. This course afforded me the opportunity to look at global trade relationships and foreign policy and their impact on household level food availability and feeding practices.

My career in nutrition started at the Feinstein International Center supporting the research of using cash, as opposed to food in complex emergencies. Following the completion of my Master's, I worked for Oxfam America in Ethiopia for seven months as an information officer, where I was responsible for working with Ethiopia NGOs to support humanitarian response to drought and conflict. I also looked at the impact of the coffee trade on food security in rural areas of Ethiopia. I then worked for GOAL for ten months as a nutrition surveillance officer, and was responsible for conducting nutrition surveys in food insecure areas all over Ethiopia. I also began working on improving the quality of the community management of acute malnutrition (CMAM) projects. After Ethiopia, I was hired by UNICEF as a consultant to conduct the Darfur-wide nutrition survey in response to the mass displacement there. I then worked briefly as a fellow on the Child Survival and Health Grants programme for USAID in the Global Health Bureau in Washington DC before returning to South

Darfur in 2007 to work as a nutrition specialist, again with UNICEF. In 2010, I was deployed by UNICEF to coordinate and initiate response to the earthquake and mega-floods in Haiti and Pakistan respectively. I then began working in the Nutrition in Emergencies Unit in a UN agency HQ, where I continue now.

In this role I support emergency response in countries ranging from the Horn of Africa and the Sahel to Syria and Yemen. In addition to helping nutritionists in regional offices to adhere to best practices, advocate for funding, and organise response, I also work on global policies and guidelines in collaboration with partners from the Global Nutrition Cluster (GNC), other UN agencies and Ministries of Health from numerous governments.

Working in emergency nutrition is both exciting and complicated; it remains a discipline where normative work is under much debate, and there is a shift to ensure that emergency nutrition response is part of a larger continuum to build resilience under the Scaling Up Nutrition (SUN) movement. Children and mothers are particularly vulnerable during crises, and improving access to quality emergency nutrition services often means the difference between life and death for children and mothers in need.

Erin Boyd (USA), Nepal, Ethiopia, Sudan, Haiti, Pakistan

Note

1   'Maternal and child undernutrition and overweight in low-income and middle-income countries', Prof. Robert E. Black MD *et al. The Lancet* – 6 June 2013.

ORPHANS AND VULNERABLE CHILDREN (OVC)

War, genocide, disease (HIV in particular) and acute poverty all have a tremendous impact on children. Included within the OVC category are those under 18 who are orphaned by the death of one or both parents; abandoned by parents; living in extreme poverty; living with a disability; affected by armed conflict; abused by parents or their carers; malnourished due to extreme poverty; HIV-positive; and those marginalized, stigmatised, or even discriminated against. They are the children who are most likely to fall through the cracks of regular programmes, or experience negative outcomes, such as loss of their education, morbidity and malnutrition, at a higher rate than their peers, with long-term impacts. Child vulnerability is a downward spiral where each shock leads to a new level of vulnerability and a host of risks. The most cost-effective interventions assist OVC before they have reached the most critical stages of vulnerability. At the bottom of the spiral are the children who live outside of family care or in situations of severe family abuse and neglect, and interventions at this level tend to be too expensive to be sustainable and have low rates of success (World Bank).[1]

The work: You will be working to protect children from negative outcomes by ensuring that OVC are given special attention either by ensuring their equal participation in projects designed to benefit all children, or through projects tailored specifically to their needs. In order not to create stigma for OVC, programming has evolved so that the target group is often wider than the OVC themselves. Targeting children at the top of the vulnerability spiral – those from a poor family or with ill parents for example – you will be minimising the impact of subsequent shocks for these children, enabling them to lead healthy and productive lives. The field is very cross-cutting and includes health, education, economic strengthening, nutrition, social protection, HIV prevention and treatment, youth programming and so on. Examples include out-of-school programmes, GBV and child protection programmes, child-sponsorship schemes, nutrition and education interventions, early childhood development programmes to give them the best start in life, and advocating for national policies that benefit the most vulnerable children.

The employers: A large range of players, including international agencies such as UNICEF, international and national NGOs with a focus on children's well-being and the rights of the child, individual orphanages and out of school centres.

Breaking in: Your entry point may be from a thematic focus such as protection, health, education or nutrition, or from a field such as social work. Working with an organisation that focuses its programmes on OVC or mainstreams this work in the field is also a good way to get a foot in the door.

Job titles: Child Protection Specialist; OVC Programme Manager

Find out more:

World Bank OVC toolkit http://info.worldbank.org/etools/docs/library/138974/toolkit/howknow/definitions.htm

See also: Protection; Education; Health professional; Nutrition

---

After a degree in Modern Languages and a Master's in Comparative European Literature, I began a PhD in African Studies but became impatient with the time the dissertation was taking as there seemed so much 'real' work to be done. So I turned my research (in African oral traditions) into a text book that is still used in secondary schools in Swaziland today. I returned to the UK and worked as a journalist for an African news magazine but the stories were often negative: failed dam programmes; corruption; etc. I used a move to the US as an opportunity to change from writing about development to actually trying to do it, hoping to create positive outcomes. I started in communications and then became a consultant for World Education – initially as a writer – and then worked on a portfolio of programmes focusing on South Africa. It was the mid 1990s and the country was moving into the era of a *new* South Africa with the release of Mandela.

With the onslaught of the HIV epidemic we began to address the impact it was having on all of our programmes in sub-Saharan Africa. But I became disillusioned, as much of the money from large funding programmes never reached the folk at community level who really needed the support. This applied especially to OVC affected by HIV and their caregivers. Much of the money in multimillion dollar programmes seemed to get stuck at mid level (international and national NGOs/district governments) and critical services and support rarely reached those who needed it most.

In 2004 I took a sabbatical in Oxford and came up with the idea for the Bantwana Initiative ('Our Children' in Zulu) to support vulnerable children affected by HIV with a strong community-based approach that supported vulnerable children with comprehensive services – starting with children and families where they were, in their communities. The OVC sector was still nascent. Comprehensive care was still a new concept and much of OVC care was siloed through health and education pro- grammes. Bantwana was launched as a new initiative at World Education, with initial funding of $50,000. The approach was timely as the sector began to grow and six years after launching this initiative, we are active in four countries (Zimbabwe, Uganda, Tanzania, Swaziland) and our annual budget is over $10 million. We pilot innovative approaches to OVC care including child protection, youth employment, out-of-school education programmes, livelihoods, savings and credit, nutrition, GBV, HIV preven- tion, children living with HIV, primary health care and psychosocial support, and once these show impact in smaller projects, we scale these models up in our larger government funded programmes.

I particularly like the programmatic aspect of my job, especially as it is not confined to one area of development and means keeping up with a broad range of sectors (health, education, livelihoods, social protection etc.). I enjoy designing new programmes together with our country teams, and when we can show an impact – improved quality of life and well-being of children and their caregivers – it is very rewarding. Unfortunately, programmes need care and feeding and without resources cannot run effectively – fundraising and proposal writing are the aspects of the work I like the least.

Gill Garb (South Africa) Zimbabwe, Uganda, Tanzania, Swaziland

---

Note

1   http://info.worldbank.org/etools/docs/library/138974/toolkit/howknow/definitions. htm (accessed 30 January 2015).

PEACEBUILDING

The peacebuilder aims to bring the sanity of peace to a world wracked by the madness of violence. Conflict is natural and, if well managed, can lead to welcome social change. But where conflict negotiation fails, it can be dramatic and brutal. Unresolved conflict can threaten the social and economic fabric of society and development efforts. The peacebuilder believes that pathways to peace exist in every conflict environment. Effective peacebuilding is multifaceted and can involve everything from high-level mediation efforts in emergencies to building positive relationships between former adversaries or addressing underlying causes of conflict before it turns violent. Disarmament, demobilisation and reintegration (DDR) is a common strategy used by peacebuilding operations when the conditions are right.

The work: Peacebuilding initiatives may be at the international, national or local level. At the *international level* you might be involved in mediation work with international organisations or diplomatic missions, including researching and advising different institutions on conflict issues and contexts. Most work on the *national level* is concentrated in conflict-affected, post-conflict and fragile states and you may be developing strategies, policies and structures to address conflict and promote peace within a state or region. *Locally*, peacebuilding initiatives focus on communities in conflict and working directly with people involved in conflict through training, dialogue or other peacebuilding interventions. You could also be working on issues of conflict sensitivity in humanitarian or development programming and integrating peacebuilding activities into these interventions in conflict areas.

The employers: Diverse actors in government, NGOs and intergovernmental organisations and civil society all work on peacebuilding initiatives. Some NGOs working in this area include Caritas Internationalis (CI), Concordis International,

International Alert, Search for Common Ground, Accord, Conciliation Resources and Interpeace among others. Most of the work will be in fragile states and a post-conflict environment.

Breaking in: You will need knowledge and understanding of conflict and its spillover effects as well as the skills to design processes and mechanisms to address conflict and promote peace. Peacebuilding interventions are often interrelated to other areas of work such as governance and democratisation, security sector interventions (DDR and security sector reform (SSR)), civil society strengthening, justice systems strengthening, human rights and protection. There are several related Master's degrees in Peace Studies or Peace and Conflict. But public administration, governance, political science, social science studies can also be relevant.

Potential job titles: Advisor on Governance and Peacebuilding; Technical Advisor for Justice and Peacebuilding

Find out more:

www.allianceforpeacebuilding.org The Alliance for Peacebuilding (AfP)
www.dmeforpeace.org Design Monitoring and Evaluation for Peacebuilding is a community of practice for DM&E peacebuilding professionals; they also advertise jobs in the area.
www.eplo.org The European Peacebuilding Liaison Office (EPLO)
www.internationalpeaceandconflict.org Peace and Collaborative Development Network (PCDN) is the leading global portal connecting over 27,200 individuals and organisations engaged in peacebuilding and collaborative development.

I earned my Bachelor's degree in Political Science specialising in International Relations at the University of Illinois. After graduating I worked for two years in fundraising at an NGO in Chicago, and then joined the Peace Corps in Morocco as a Small Business Development Volunteer working with local cooperatives. I decided I wanted to specialise in conflict resolution and peacebuilding as I saw the incredible impact conflict can have on people's lives. I enjoyed development work, but I realised that peace is a prerequisite for sustainable human development.

I applied for graduate programmes in International Relations and was accepted into the Master of Arts in Law and Diplomacy at the Fletcher School at Tufts University, a programme that I chose because of the flexible curriculum, practitioner orientation and international focus in the classroom and the student body. Prior to beginning the degree I interned at Search for Common Ground in Washington, DC for three months to get more exposure to the field. During my two year Master's degree I focused on international negotiation, conflict resolution and security studies and went to South Sudan to complete research for my Master's thesis. Here I worked at a government institution focusing on community security and small arms control.

I also interned part time with the International Center for Transitional Justice, assisting with a project to bolster their internal programme design, monitoring and evaluation capacity.

In my last semester I worked with an organisation on a project focusing on governance and peacebuilding in Sudan and South Sudan and once I graduated I continued with the organisation based in Boston but travelling often to the field. I then decided to relocate to the field and took a job with Concordis as the Country Director based in Juba, South Sudan. As the Country Director I oversee all aspects of the organisation in South Sudan including programming, finance, HR, administration and liaison with donors, partners, government and others. I most enjoy developing the strategic trajectory of the programming, which requires an in-depth knowledge of conflict in the area and thinking creatively about what types of interventions can encourage peace. I lead our programme team to continually assess the dynamics of conflict, track conflict incidents and trends, and understand the various actors and their roles. From there we must prioritise how we can most effectively encourage dialogue and reconciliation. It's often difficult, as many of the issues appear intractable and change constantly but working at many levels within society (international, national, state and local) usually means that there is an opportunity somewhere to make a difference.

If you want to work in the area of peacebuilding, gaining in-depth experience and knowledge of a particular conflict area can help you to begin to understand the complex relationships between conflict and peacebuilding (and other) interventions. After gaining some experience, specialised courses in mediation, facilitation, specialty issue areas or other related fields are very important for continued growth in your career. Because peacebuilding is still a developing field, staying up to date with the latest academic and practice-oriented literature is necessary, as is meeting with other people in the field to discuss best practices, methodology and other conflict contexts. It is also important to take personal time and have a balanced life to avoid burn-out or becoming too mired in the difficulties and harsh realities that working in conflict exposes one to.

Mandy Gardner (USA) South Sudan, Morocco

POLICY

The world is full of policies. A policy is a statement of intent – a principle to guide decisions and achieve rational outcomes. Policies can be guidelines, rules, regulations, laws or principles. Governments have policies, but do not always enforce them. Development and humanitarian organisations have policies: clear and simple statements of how that organisation intends to conduct its services, actions or business. Donors have policies highlighting what their development priorities are, what their activities will fund. Humanitarian organisations have policies on how their staff will be protected, when and which philosophy will govern their interventions. Businesses have policies that guide their operations – how they buy, sell and manufacture. Policies are not static and by using the right strategies those working in the field can be successful in influencing all aspects of policy. Creating a vision is what drives those in the policy field.

The work: Policy professionals create, research and amend legislation and programmes that affect the development and humanitarian arenas in a variety of areas, from the economics of development aid – where it should be channelled and why – to how services are implemented and who they target. A policy officer will ensure those affected by the policy abide by it, and inform people of relevant policies. A policy analyst is involved in the evaluation and formulation of policy, either within the organisation or targeting other external organisations with influence. They ask what has worked, what hasn't and why, to adapt the policies and legislation accordingly. Policy advisors provide strategic advice to key decision makers, encompassing both strong analysis and nuanced understanding of the political context that can drive decisions. A policymaker will help to formulate their organisation's policies, have them approved and implemented.

The employers: You could work for a think tank or a policy research organisation who sets out to influence *policy* ideas and decisions. You could work within larger

NGOs, influencing internal policies or external actors. Bilateral and multilateral donors also have significant policy researchers.

Breaking in: There are a number of different undergraduate degrees that cover public policy areas such as comparative policy. Postgraduate degrees allow you to specialise in particular areas. Foreign or global policy are common specialisations, with areas such as peace policy often offered as a separate specialisation. But, equally, a background in research and an ability to understand and interpret data is important. Knowledge of decision-making processes and incentives for key actors such as businesses, governments or multilateral organisations can also be as important as knowledge of the thematic subject matter.

Job titles: Policy Officer; Policy Analyst; Policy Advisor; Policymaker

Find out more:

H. Jones *et al*. (2012) *Knowledge, Policy and Power in International Development: A Practical Guide*. Bristol: Policy Press Publications.

See also: Advocacy and campaigning

I did undergraduate degrees in Business (Finance) at the University of New South Wales and Law at the University of Sydney. I then went on to spend a year researching the impacts of trade liberalisation at the Australian National University and later completed a Master's in International Law and International Relations. Throughout my studies, I worked as a trainee corporate solicitor and ran a start-up furniture business in Australia. Following my studies, I effectively started my career as a Market Strategist at Procter & Gamble based in Sydney, Australia for two years before making a career move to development. After volunteering at a refugee camp in Ghana, my first development job was in working for the Australian government (AusAID) on market-based aid programmes in Indonesia.

In 2011 I joined Oxfam in the UK, working as a Private Sector Advisor in the Oxford head office. My role encompasses working across the organisation's programme and campaigning priorities. I specialise in the links between business and poverty, with a focus on agriculture and the rural economy. I advise policymakers, businesses, campaigners and programme managers on how to leverage business and markets to better tackle poverty, pushing the business sector to take tangible steps in order to have a more positive force on the lives of those living in poverty. This means I'm attempting to influence decision makers (often policymakers) rather than being in charge of policy myself. It requires me to understand the policy options decision makers have and the incentives that are driving these decisions. At Oxfam, we can campaign, programme or lobby to effect change, and it's my job to ensure we use the right tools and push for the right changes.

I am particularly interested in policy work because the right policies can impact the lives of millions, channelling inclusive investment, improving access to health and education, and promoting democracy and justice. I'm a macro kind of guy and wanted to work on generating impact on a greater scale, even if it's more abstract and less direct than alternative paths in development.

A typical day could involve meeting with business leaders to share our insights on how their companies are impacting poverty, assisting Oxfam programmes that are trying to help farmers make better use of markets, or designing campaigns to draw attention to business behaviour that is impacting those living in poverty. This means I am sometimes in Kenya working to understand how farmers and workers are impacted by global business and sometimes in London talking to businesses about what they could be doing differently.

I like being able to see powerful institutions change their policies in a way that improves the lives of real people. When we saw the world's biggest food companies start to require their suppliers to give communities a voice in any land acquisitions, the danger of land-grabbing subsided in many communities around the world. This was the outcome we had worked for for so long through our Behind the Brands campaign, which used a combination of advocacy and campaigning tools. But the key objective was to achieve policy change, starting with the world's largest companies.

If you want to get into this field, build knowledge of how various actors make policies. Whether it's business, government or NGOs, try to get experience understanding incentives and processes in multiple contexts. It's as much about knowing the players as it is about knowing the topic. Follow key people on Twitter and know the thought leaders who drive the debates and influence key decisions.

Erinch Sahan (Australia) Indonesia, Ghana, Kenya,
Azerbaijan, Georgia

PRIVATE SECTOR DEVELOPMENT

A market-based approach to poverty reduction, such as business and job creation – especially focusing on youth employment – is a core strategy for economic development in many countries. Moving away from donations and direct support – which can create dependency – private sector development (PSD) aims to strengthen the private sector, enabling poor people to be active players in the economy either directly (through entrepreneurship and business creation) or indirectly (through enhanced employment opportunities). Economic revitalisation, the promotion of innovation, in particular technological innovation and *green growth* have become fashionable subjects in this field. The focus may be on supporting the start-up and growth of micro, small and medium enterprises (MSMEs) and helping them to overcome some of the challenges that they face. These can include the lack of collateral to access loans and the low levels of education and managerial capacity found in many developing countries. PSD, economic development and poverty reduction are often closely linked concepts.

The work: You will most likely be involved in helping to create an enabling environment and reducing barriers to doing business. This could involve working with banks to provide loans and credit to emerging businesses or establishing guarantee funds; coaching entrepreneurs in business plan development; supporting business incubation centres; advising governments on how to reduce red tape for new businesses; running business plan competitions; value chain development; ensuring that women, youth and marginalised groups with no collateral and little education have equal opportunities in business and access to finance; strengthening exports, linking entrepreneurs to markets, product diversification, promoting export and trade competitiveness and productivity. Other jobs with this same title might be something quite different: enhancing and influencing an organisation's

306 Areas of speciality

own links with private sector actors to increase their involvement in development, focusing on partnerships and advocacy.

The employers: This is a core focus of some of the major multi- and bi-lateral donors (World Bank, IFC, GIZ, DFID, USAID) whose projects are often implemented by consulting firms employing external experts to deliver this work. Several NGOs and universities also support the work in this area. The ILO is a big player in this field.

Breaking in: An MBA, Business and Development or Finance advanced degree will be a definite advantage. Knowledge of advocacy and policy is essential as is expert knowledge of the role of the private sector in development.

Potential job titles: Business Training Specialist; Private Sector Advisor; SME Business Development Expert; Private Enterprise Development and Access to Finance Coordinator

Find out more: The Donor Committee on Enterprise Development (DCED) has become the leading source of knowledge about PSD – documenting and disseminating the successes and lessons learned to improve the results of PSD programmes in developing countries. www.enterprise-development.org

The Global Partnership for Youth Employment (GPYE) goal is to build and disseminate evidence on youth employment outcomes and effective programs. www.gpye.org

Lael Brainard (2006) *Transforming the Development Landscape: The Role of the Private Sector*, Washington, DC: Brookings Institution.

C. K. Prahalad (2006) *Fortune at the Bottom of the Pyramid: Eradicating Poverty Through Profits*, Upper Saddle River, NJ: Dorling Kindersley.

See also: Microfinance; Value chain; Governance; Law and development

---

After finishing high school I travelled a lot and envisaged myself as an international business person, so decided to aim my academic studies towards international economics. I started with Social/Human Geography at Utrecht University, and then moved on to an MA in International Economics and Economic Geography. Here I was introduced to the world of business in developing countries.

After university I tried to get into various UN employment programmes focusing on private sector development, and ended up doing an internship at UNDP Honduras – on cross-border economic development strategies for Central America. After this I started to work in Bosnia-Herzegovina as a programme manager for a small youth development NGO. Although the work wasn't directly related to where I wanted to be, the experience brought me many insights to the world of development. After some years I went back to private sector/economic development and co-started a couple of social enterprises alongside consultancy work in the Western Balkans, eventually to be picked up by a global consultancy firm back in the Netherlands to work on projects to enhance Business Development Services and access to markets. After three years of running (small) projects, I changed to my current employer, working on a large, long-term, multi-country enterprise development programme. Combining my expertise in a management role has been a welcome development.

In my current job I oversee the programme in five countries, liaising on a regular basis with the Country Managers. We support start-up and growth entrepreneurs through various trainings and programmes, including business plan competitions. For example, we have been working with banks in Liberia and Burundi to set up a loan guarantee fund, enhancing access to finance for competition winners. In Kosovo and Liberia we also set up Business Start-up Centres from where our programmes are run, and we are starting an entrepreneurship academy to stimulate generation of new innovative business ideas. Our work focuses on post-conflict countries as we see job creation as a way to empower ambitious youth to lead societies to prosperity. As an NGO working in the field of PSD, one of our main considerations is ensuring that our support is not distorting the local market. We facilitate people to access commercial loans and create bankable business plans, without becoming part of the system. I enjoy working with dynamic people who don't rely on hand-outs but want to lead their own development.

A major part of working in development is quality standards, which is essential for our work towards partners and donors. The consequence is a large documentation requirement from us and our partners that needs a lot of coordination and checking. For those wanting to work in the field, it's a good idea to create a good and consistent 'red line' in your CV. Higher education is essential, especially for larger INGOs, and preferably with good grades. This is a difficult market to get into – lots of education and internships are good to do.

Marc de Klerk (Netherlands) Balkans,
India, Vietnam, Liberia, Ghana.

PROCUREMENT

The procurement manager buys for a living. Many development and humanitarian projects require the purchase and supply of equipment, drugs and consumables that often amount to millions of dollars in value. Not only is the purchase and supply of these products vital for project implementation, but these orders can also make a significant economic impact on the companies chosen to supply the products. Those working in procurement have the opportunity to influence this. For example, if grain for food aid is sourced from African farmers rather than from the resource-rich countries, it has the potential to transform the local economy. Companies purchasing from developing countries – even if not directly involved in development projects – also have the power to contribute to poverty reduction. When successful, such trade can provide new opportunities for wealth creation and a route out of poverty for those concerned. The procurement team needs to balance value for money alongside responsible procurement, giving weaker economic players such as the small-scale suppliers an equal opportunity.

The work: The procurement manager ensures the efficient management of procurement processes. This might involve obtaining quotes for commodity purchases; and transportation in accordance with appropriate donor government regulations; negotiating contracts to get best value for resources; preparing bid evaluation forms for field and/or donor review; preparing purchase orders and contracts; ensuring that the products supplied meet appropriate quality standards. You might also be involved in the elaboration and implementation of a sourcing strategy and e-procurement tools. You might lead efforts in the procurement, shipping, installation and commissioning of technical equipment and spare parts as well as being involved in designing bidding documents and the tending process.

You will ensure that project procurement and subcontract procurement activities are within budget and in compliance with the prime contract. Procurement managers often work closely with the logistics team.

The employers: Any organisation that is involved in the purchase and supply of goods in order to implement a project. This might be for an NGO, a consultancy company and some multilateral organisations implementing projects. It could be in an emergency context (such as food aid, tents) or in a development project (health care products, books, engineering equipment). You will work closely with the logistics team.

Breaking in: It usually starts as learning on the job, or through a training programme, complemented later by academic studies. Commercial and business skills, alongside knowledge of the specific field you are working in (be it health, education, emergencies) will be an advantage. Knowledge of trade and development will also be an asset.

Job titles: Procurement Assistant/Officer/Specialist/Expert; Purchasing and Supplies Manager

Find out more:

Humanitarian Procurement Guidelines: http://ec.europa.eu/echo/files/partners/humanitarian_aid/Procurement_Guidelines_en.pdf

See also: Logistics; Food security; Humanitarian response management

---

I stumbled into procurement as a career by accident. I had been a banker for several years in Nigeria and after that worked as Head of Personnel and Admin in a privately owned radio station in Nigeria. I have a very commercial background. I took a gamble when I responded to an advert for an Assistant Commercial Manager. At the time I applied I did not have any procurement experience but believed that my commercial background would stand me in good stead. It paid off and they engaged me as a Trainee Procurement Officer.

My interest in procurement blossomed when I realised that this was an area where I could make a significant and positive impact in my country both for economic development and in ensuring the provision of necessary goods and services to Nigerians living below the poverty line. Value for money procurement can lead to huge cost savings and sourcing for products and equipment locally, can bring about the development of commerce and industry which could lead to expansion of the economy. We source for drugs, consumables, IT equipment, office equipment and furniture locally but sometimes we need to source internationally for medical equipment, machinery or high-value vehicles.

I am now Senior Procurement Manager for Crown Agents Nigeria, a Development Consultancy, and I have been involved in the procurement of over $20 million worth of anti-retroviral drugs, drugs for opportunistic infections, medical and lab equipment for the National Agency for the Control of HIV AIDS (NACA) and the Society for Family

Health funded by Global Funds (Round 9) to fight HIV AIDS. I have also been instrumental in the procurement of about £12 million worth of drugs, consumables and medical equipment for health facilities across Nigeria, which is funded by the UK DFID under the Partnership for Transforming Health Systems 2 (PATHS) programme for health facilities in Nigeria. This is in addition to several other high-value health and other non-health programmes working in Nigeria.

My typical day starts with setting of targets for the day or for the week. I check that orders are on track, authorise tender documents, contracts or payment requests, call for meetings with suppliers to resolve knotty issues. I also update the various procurement/financial status reports for the different programmes we work for, and monitor all team activities. It is a high-pressure job that demands the ability to think on your feet and deliver results within tight timelines and in a developing economy; however, I enjoy this work because I tend to excel in challenging roles.

Continuing professional development in procurement is necessary to stay up to date with new and emerging trends and practices in the procurement landscape. I am currently undertaking professional studies to obtain my Graduate Diploma in Procurement and Supply from the Chartered Institute of Purchasing & Supply UK (CIPS).

Sandra Odogwu, (Nigeria)

PROJECT MANAGEMENT

Whether focusing on agribusiness, health, education, employment or affordable housing, what many development workers have in common is that they try to change the world through projects. A project is a temporary endeavour to create a unique product, service or result.[1] But the project results depend on the ability of project managers to deliver the project efficiently and effectively, and this is anything but easy. Within the development sector, operating environments are complex, challenges numerous, relationships complicated and the cost of failure is high. To succeed, the project manager must proactively and decisively manage these challenges.[2] Many development projects fail to deliver on scope, budget and time frame and this is largely due to poor or unrealistic project design and the inability of the project manager to identify challenges that can impact on the project and manage them accordingly.

The work: The successful project manager is an all-rounder combining both soft skills (enabling them to lead, enable, motivate and communicate) and hard skills, following who-does-what-when (planning, estimating, measuring and controlling the work). The project manager will often be involved in all elements of the project cycle, from project conception and design, to sourcing funding, implementation and continuous monitoring and evaluation of the work. They do not implement the project themselves, but manage inputs – be it personnel or resources – and outputs. They will need to have up-to-date detailed knowledge about each element of the project, juggling the day-to-day activities with a longer-term strategic view, constantly monitoring progress and making necessary adjustments to activity plans and budgets. Programme management is the process of managing a group of related projects in a coordinated way. Projects can be very small scale of only a couple of thousand dollars funding, to large-scale multi-donor, multi-annual and multi-country contracts of several millions.

The employers: Almost all development and humanitarian organisations including NGOs, consultancies, research institutes, governments and multilateral agencies run projects and programmes, and thus need project/programme managers for effective implementation.

Breaking in: People may come to project administration or coordination at entry level, with little training or knowledge of the field. A lot can be learned on the job, but it is also strongly advisable to get some specific training – there are many short courses available; for example, the Project Management Institute (PMI) offers specific training courses and project management software. Knowledge of specific project management software can also be relevant for larger projects and programmes. Specific Master's programmes in Project Management also exist.

Job titles: Project Coordinator; Programme Officer; Programme Assistant; Programme Manager; Programme Specialist; Programme Director

Find out more: Project Management for Development Organisations www.pm4dev. com/ offers resources and short online courses for professionals working in development; 'The Fundamental Project Management Course' for example runs several times a year and costs around $100.

'A Guide to the Project Management Body of Knowledge', the now-in-development ISO standard for project management, http://books.google.rw/ books/about/A_Guide_to_the_Project_Management_Body_o.html?id=FpatMQ EACAAJ&redir_esc=y

---

I studied languages and then did a Master's in Information and Communication Sciences in Senegal. This led me on to my first job as Information Assistant in 1997. I then joined Oxfam in 2002 as a Regional Programme Information Officer. Here my role was to support countries and regional programmes in their management, monitoring and evaluation efforts. This experience brought me close to various communities, partner organisations, government services, campaigners and donors, which gave me a strong foundation from which to work as I had senior programme management responsibilities. I came to realise that communication and effective learning are vital factors for successful programme management. My employer offered me wonderful opportunities to receive training in many aspects of project management but my commitment and willingness to learn from successes and failures have been critical to my career development.

In 2007 I moved to Niamey in Niger to take up a role as Programme Manager. I worked there on an education programme but was also engaged in all areas of programme management: financial management, monitoring and evaluation, partnership, fundraising, and deputising for the Country Director whenever he was absent. As a Programme Manager, you have a lot of responsibility and are under a lot of stress but you can also enjoy the space and power you have. One of the challenges of managing a programme is that you are not necessarily a specialist in that area. But it is up to you to learn as much as you can about the field, and surround yourself with experts who can inform and advise you. Your job is to ensure that the programme runs smoothly, efficiently, delivers what it set out to do, but is also flexible enough to change if necessary. There should be no bottlenecks that delay or negatively affect the programme. I did not specifically learn project management at school but on the ground and through intensive and practical in-house training.

In 2009 I went on to become Country Director of the Oxfam GB office in Niger. This was a different role from Programme Manager in that I became the most senior and visible member of the team. My scope of work went beyond single department/ sector management responsibility. The change was significant and called for other skill types including leadership, listening, patience, communication and people management. But my previous experience in programme management helped me to make the change and drive the programme successfully. In 2012 I was very proud to receive an MBE, Honorary Member of the British Empire, from Queen Elizabeth II, an award that I received for my services to Oxfam in Niger during the food crisis.

I am now Regional Programme Manager and I oversee the work of several country programmes, most of which are within livelihood, governance, resilience, human-itarian and advocacy fields. My advice to anyone working in project management, or wanting to get into it, is to combine different learning options: read, listen, talk to people, observe and be patient. A good project manager is a field worker, a facilitator, an educator, a guide, a leader but also a good learner.

Mbacke Niang (Senegal) Mali, Niger

Notes

1   *A Guide to the Project Management Body of Knowledge* (PMBOK Guide) Fifth edition, Project. Project Management Institute, 2013.
2   *A Guide to the PMD Pro: Project Management for Development Professionals.* Developed by APMG-International in partnership with Project Management for NGOs (PM4NGOs).

PROPOSAL WRITING

Grant and proposal writers research, develop and author project proposals. These will respond to a donor's (e.g. USAID, Comic Relief) call for proposals or a tender, and therefore will need to comply with their exact specifications. Proposal writers work closely with regional and technical experts to create relevant projects with proposals written in a coherent, clear and convincing way. The ability to conceptualise, interpret and synthesise large amounts of technical information and translate complicated ideas into easily understood language is important. In order to be effective as a grant writer, it is also important to understand how development projects are designed and the strategies that are used by organisations to implement projects. Proposal writing is a core component of development work and is a highly desirable skill – it is, after all, the process that drives funding for so many meaningful projects. In many cases project managers, fundraisers or chief executives will lead the proposal writing process, while larger firms may have designated technical writers or hire consultants to complete this work.

The work: A funding round or call for proposals initiated by a donor is identified. If the eligibility criteria and focus is in line with an implementing agency's remit, they will decide to submit a proposal. The first stage is usually to submit a concept note of the work. If this is approved, a full application will need to be developed. In order to develop the project concept, draft the narrative section of the proposal and develop the logical framework and detailed budget, the proposal writer may need to travel to the field to meet with technical experts, gather background research, or conduct a needs assessment. The proposal writer will then integrate information and technical inputs from these different sources into a well-structured proposal. Finally, they will need to edit and format all proposals for readability, consistency, and ensure they adhere to client and donor specifications. The tendering process for some donors requires details of the team implementing the project to be included at the proposal stage, even before funding is secured, so finding the right experts or partners, negotiating partner roles, or liaising with HR may also be a central function.

The employers: Grant writing is a core component of implementing agencies' work (consultancies, NGOs, UN agencies, etc.) as it is the main way to access restricted project funding from a donor or win a contract. Most people working in these organisations will be involved to some extent in developing funding proposals – whether it is writing them or through providing guidance – even if it is not a core element of their role. Some organisations hire full-time proposal writing staff, while others outsource to consultants.

Breaking in: Clear concise writing is essential. As proposal writing is such a core component of many organisations' work, getting exposure to this is relatively easy. It's an important skill to have for many jobs. If you want to specialise in this area, technical expertise relating to the area of the proposals is a distinct advantage, as is experience implementing or managing development/humanitarian-related projects. Familiarity with the donor proposal process, requirements and a proven track record of producing winning proposals is important.

Job titles: Technical Writer; Proposal Writer; Grant Writer; Lead Author

Find out more: www.euroresources.org/ publishes a *Tips and Tricks* to access funding from a variety of resources.

My Ugandan heritage fuelled my interest in international development. Travelling and working internationally have helped me to gain rich insight into a part of the world where I felt a deep connection. I complemented a sociology degree with courses in international relations. After working for a few years, I took the leap and moved to Uganda where I volunteered as a Research Assistant on a study examining migration patterns of health care professionals. A year later, I pursued a Master's in Diplomacy and International Relations and secured a consultancy with UNESCO in Paris, where I worked in the Communication and Information Sector. This provided my first opportunity to write about international projects and broadened my understanding of the complexity of working with multilateral organisations – indispensable for my future career.

The trajectory that I followed to a career in proposal writing was neither straight nor deliberate; rather, it built on the set of skills and experiences I gained through various opportunities. I was hired as a Project Coordinator for John Snow, Inc. (JSI), a public health firm headquartered in Boston. It was here that I first gained exposure to the process of proposal development and writing, as part of my role coordinating integrated health and maternal and reproductive health projects in Egypt, Djibouti and Ukraine. Supporting proposals was a core aspect of the job and I was involved in nearly all aspects of the process. I later joined World Education, Inc. as the Country Director of their portfolio of projects for orphans and vulnerable children in Uganda and assumed an even greater role in the development of proposals.

Since I left World Education, Inc. in 2009, I have been working as a consultant, specialising in research and technical writing for international programmes. I am mainly hired by organisations as either a Technical Writer – where I am one of two or three writers on a proposal team, with each writer leading the research and writing about a particular section or component – or a Lead Writer/Author where I am responsible for the entire technical narrative. I focus on the thematic areas of HIV and AIDS, social protection, education, and vulnerable children and youth.

Aside from the more obvious skills in writing and editing, the process of proposal writing requires tremendous flexibility. Writing often occurs in tandem with project design and thus the document is continually being refined, based on emerging feedback from team members. This all has to be seamlessly integrated into a cohesive and persuasive document. You should have strong communication and project management skills to keep the process moving and on track. The work also often involves long hours and tight deadlines. But when a grant is awarded, offering the potential for positive outcomes in the lives of the beneficiaries, it's all worthwhile.

My advice for those pursuing a career as a proposal writer would be to gain as much technical knowledge as you can as you are often called on to provide technical expertise in addition to writing skills. Build up skills and expertise where you can.

Nansubuga Mubirumusoke (USA) Uganda,
Tanzania, Zimbabwe, Zambia

PROTECTION

The Inter-Agency Standing Committee defines protection as: 'all activities aimed at obtaining full respect for the rights of the individual in accordance with the letter and the spirit of the relevant bodies of law (i.e. HR law, International Humnitarian Law (IHL), refugee law).' Those working in this field focus on the protection of civilians during armed conflict, or a humanitarian emergency. The definition is very broad, so complementarity and collaboration between agencies is a key component in a humanitarian context. While every context is unique, there are protection concerns that occur with depressing predictability, and demographic groups – children, women, those who are frail, whether through illness or advanced age, socially marginalised minorities – tend to be particularly vulnerable in times of crisis. For this reason, the protection sector has spawned certain sub-groups to focus on these issues. The list may vary according to the threats extant in a given situation, but Child Protection and SGBV are two that are usually represented.

The work: Given that protection touches on virtually every aspect of humanitarian work, you could be involved in an incredibly diverse range of activities. You might be putting a stop to a specific pattern of abuse or alleviating its immediate effects, restoring people's dignity and ensuring adequate living conditions through reparation, restitution and rehabilitation, or fostering an environment conducive to the respect for the rights of individuals. Protection activities may include responsive action, remedial action and environment building, and may be carried out concurrently. Protection officers might work at the 'corporate' level within their own organisations, developing the overall position on protection and its mainstreaming. There might be training to conduct for staff and implementing partners or helping authorities develop their capacity to meet their protection obligations by advising in policy development or even legal reform. At field level you might operate emergency services for victims of rape. The scope is enormous.

The employers: Most organisations working in post-conflict or humanitarian emergency situations will have activities that fall under the protection banner. UNICEF is the lead UN agency for Child Protection and NGOs such as Save the Children have a child-protection mandate. The Heartland Alliance works on lesbian, gay, bisexual and transgender (LGBT) rights and protection in developing countries.

Breaking in: An academic background in law or human rights is an advantage. A background in early childhood development, youth affairs or education can be relevant for people wanting to specialise in child protection. Likewise, home-country experience in social work, particularly work related to sexual or gender-based violence will be directly relevant, given the prevalence of this problem in most emergency settings. Save the Children UK have an annual Child Protection Trainee Scheme.

Potential job titles: Protection Officer; Child Protection Officer; SGBV Programme Advisor, Protection Cluster Coordinator

Find out more: The Global Protection Cluster www.globalprotectioncluster.org is the best starting point, with links to resources, vacancy notices, sub-cluster websites and interagency meetings/consultations etc.

See also: Human rights; OVC

I had a tendency to climb up onto my social justice soapbox whenever the chance arose, but the idea that this could actually be the basis of a career wasn't initially obvious to me. So, after school, I went into journalism. As soon as I'd finished my cadetship I set off on a two-year period of travel that sparked daydreams of working as a foreign correspondent, and ultimately sent me back to university to study Arabic, and then to Beirut. But by then it felt as if the media was losing its soul and I started thinking about what I might find more fulfilling.

While I'd been travelling I had met people who shared their experiences of discrimination, persecution and living with conflict. Their stories resonated with me. Around the same time, Australia had instituted a harmful mandatory detention policy for asylum seekers arriving by sea and as I followed these issues I was exposed more and more to the work of advocacy groups. I got involved with providing support to newly arrived refugees in a voluntary capacity initially, and was active with Amnesty International as well. At the same time, I was completing a Master's degree focusing on human rights. Not long afterwards I began volunteering and later working for Australian Red Cross (ARC), initially on International Humanitarian Law, and then in the International Tracing, Refugee and Asylum-Seeker Services department.

I entered ARC's international delegates' roster and was posted to the Pacific region – Fiji, then Papua New Guinea. Later, I did a series of consultancies with the IFRC secretariat, focusing on migration and protection. Then, wanting to get back out to the field, I joined RedR Australia's standby roster, which has given me the opportunity to do missions with several UN agencies (OCHA, UNHCR, WFP and now UNICEF) in Darfur, Afghanistan, Haiti and Jordan.

As I've moved from organisation to organisation, the core elements of my job have been distinctly different. There have been organisations focused squarely on protection, such as UNHCR and UNICEF, and non-protection mandated organisations, that nonetheless want to incorporate an understanding of protection into the way they approach their core business. I've worked on 'mainstreaming', which involves advocacy within the organisation on the relevance of protection to all activities, and training to give staff the understanding and skills to apply a 'protection lens' to how they approach their work. At other times I have worked on programmes to deliver services to IDPs, asylum seekers and refugees, or in roles that routinely required me to find solutions to problems that really should not have been routine, such as women or children being deported into highly insecure environments without resources or support networks.

Protection work has its challenges. For a start, there are actors who don't accept human rights as a legitimate basis for humanitarian action and insist that humanitarian response should be limited to providing material assistance to save lives. There is, however, a very real satisfaction knowing that you have helped to either make the system work better, or at least helped people evade the worst impacts of its breakdown.

Robyn Kerrison (Australia) Fiji, Sudan (Darfur),
Afghanistan, Haiti, Jordan

PUBLIC HEALTH

Public health focuses on the preventative rather than the curative side of health. It is a multidisciplinary field, looking at the health of a population as a whole and encompasses social, cultural, economic and political aspects of health promotion and prevention. At the macro level the focus may be on health systems strengthening programmes – addressing all the pillars of an effective health system, in recognition that without trained staff, health financing, management and appropriate policies in place, health outcomes are unlikely to be met. You could also be working on a specific disease or issue-specific areas such as HIV and AIDS, reproductive health or immunisation campaigns. Public health practitioners work both in development and emergency contexts, but the approach can be quite different. In the latter there is a strong emphasis on rapid health appraisals and reliance on standard guidelines. In development contexts the work has a longer-term focus, and brings together decision makers, health practitioners, citizens and civil society to plan and implement public health programmes.

The work: Public health practitioners can play a very diverse and interesting range of roles in health education, biostatistics, environmental health, epidemiology, health services administration and management, maternal and child health, nutrition, and public health policy and practice. Promoting the use of bednets to prevent malaria is a good example of a public health programme as is the elimination of mother to child transmission (eMTCT) in HIV. As a public health practitioner you may be working on the front line, designing and implementing behaviour change programmes – for example, working closely with the communities or focusing on policy, capacity building or programme management.

The employers: As a cornerstone of development, a large number of organisations exist that employ public health professionals in both development and emergency contexts. The WHO leads the UN policy and health guidelines, The Global Fund to fight TB, Malaria and HIV funds most of the HIV work across the globe at a national health system level. There are many NGOs, development consultancies and other bi- and multilateral organisations working in the field of public health in both development and emergency contexts.

Breaking in: Public health is broad and interdisciplinary and as such, there is no set entry point or standard background to get involved in this field. You can enter public health from a political, health systems, clinical (e.g. medical, nursing, midwifery), social or communications background. A Master's in Public Health (MPH) will certainly help you advance your career. Opt for an MPH with an international focus or gear your studies towards this area, particularly if you want to work outside your own country of practice. In the UK, the London School of Hygiene and Tropical Medicine (LSHTM) or the Liverpool School of Tropical Medicine (LSTM) run such programmes.

Job titles: Public Health Consultant; Health Systems Strengthening Specialist; Community Health Expert; Behaviour Change Communication Technical Advisor; Public Health Advisor

Find out more: www.jobspublichealth.com/topjobs.html often lists senior public health jobs

See also: Maternal, child and reproductive health; Health professional

---

My undergraduate degree was in German and Business Studies and I then worked in an investment bank. After a few years I realised I wanted to do something more socially oriented and spent around a year looking for alternative jobs. I eventually got a job with a local NGO in Edinburgh, supporting people with epilepsy in a role as Training and Public Awareness Officer. I stayed there for three years and found it fascinating. I did a lot of work around stigma reduction, training, working with communities, schools, changing attitudes towards epilepsy. I realised I wanted to work within the health field, but it was too late to become a doctor, plus I really loved the prevention and community side.

I knew the London School of Hygiene and Tropical Medicine was well respected in this field, so I chose to do a Master's there in Health Promotion which focuses on prevention and behaviour change. After graduating I wanted to stay in London as it has such a diverse mix of ethnic groups and I thought it would be interesting to work in that context. I worked on a child obesity prevention programme in Southwark. I took a three-month sabbatical from this job to do an internship with UNHCR in Ghana. I was working in a refugee settlement on an information monitoring system – not what I wanted to do but what they needed, and it helped me by giving me some international experience.

I then took up a job in East Africa with GIZ. It was a Sexual Health Advisor position, setting up youth-friendly services to prevent teenage pregnancies and STIs. From there I took up a job in eastern Congo with IMC on a large USAID funded programme. It was a behaviour change communication programme addressing GBV. We covered two provinces, North and South Kivu, with six sites of operation, two urban and four rural. The programme I managed, with 14 local staff, worked with communities to change attitudes and practices linked to violence and gender norms. We addressed attitudes and practices towards women and worked with communities as the change had to come from them. We did local community mobilisation activities, promotion of positive role models as agents of change and creation of GBV prevention coalitions. There was a mass communication element, using radio, posters, leaflets, etc. Eastern DRC is a war zone, with a lot of war-related rape. So we worked with the military in addition to other stakeholders, such as the Ministry of Gender, local organisations and UN agencies who head up the coordination system for sexual violence. IMC also had a sister programme, focusing on the response: medical, legal, psychosocial and livelihoods assistance.

The job was 50:50 in the office and in the field. I oversaw the management and admin tasks of the programme, in addition to ensuring our methods and approaches were technically sound from a health promotion perspective, and training, supervising and supporting the staff in the field. Change is slow but you do see some small outcomes, which is encouraging; for example, when the community gets engaged, and takes the messages on board. The challenge from donors is that they want to see immediate results, while behaviour change is slow and takes time.

Alessia Radice (Italy) DRC, Rwanda, Ghana

RESEARCH AND ACADEMIA

The purpose of research is to generate knowledge to inform action. Research thus forms the basis for programme and policy development. For the fight against poverty to be successful, work needs to be guided by a sound evidence base. What works, what doesn't, and why? The first challenge is defining problems and asking the right questions: contextualising research in the larger body of evidence, and producing knowledge that is applicable outside of the research setting with implications that go beyond the group that has participated in the research. One of the challenges is the slow transition of research into policy and practice. There is often a disconnect between those who create the evidence base and those who are positioned to implement the research findings. But increasingly researcher, policy makers and planners are working more closely together.

The work: You will be helping to construct knowledge to promote global change and improve the effectiveness of development and humanitarian initiatives. From identifying research questions, developing the research methodology and parameters of the study, to actual implementation, analysis of the findings, writing up, and presentation and facilitating the implications to be transferred to practical policy change. If you are based at an academic institution you might also be involved in teaching and supervising students, building the next generation of researchers in the field. At any one time you will most likely be working on a book, or planning your next one, writing some research papers in addition to your teaching responsibilities. Some projects may be very short term, or in response to the information needs of a particular organisation that sponsors the research, while other studies may be much longer term, bringing together whole teams of people.

The employers: The academic job market is quite competitive, but research jobs don't need to be confined to universities. Many bilateral and multilateral organisations, research think tanks and larger NGOs also do similar work. While you can go in to research with a Master's, a PhD will prepare you much better for designing and conducting research.

Breaking in: A PhD is not essential for research jobs, but it is advisable as it will give you a much better foundation for them as a PhD prepares you for designing and conducting research. The usual academic route is Master's, followed by a PhD and a post-doctoral position. Having some practical experience, either with NGOs, international organisations, or multilateral agencies is useful before embarking on a career in academia as it will help you to set the context.

Job titles: Researcher, Research Officer, Research Associate, PhD Candidate, Post-doctoral Researcher, Assistant Professor, Associate Professor, Professor, Fellow, Lecturer, Senior Lecturer

Find out more:

Sophie Laws, Caroline Harper, Nicola Jones, Rachel Marcus (eds) (2013) *Research for Development: A Practical Guide*, London: Sage.

See also: Knowledge–policy interface; Policy

My interest at school was in writing, so I worked at a local newspaper in Pakistan where I was given an assignment to write on different NGOs in the region. That is what got me interested in development, but at the time there were few development specific courses (probably a good thing) so I did a double major in Economics and Political Sciences at McGill University (Canada). After graduating I came back to Pakistan and I was trying to figure out what my interest in development meant to me from a career perspective.

I started working with an NGO in Lahore with low-income neighbourhoods and slum development but realised that this wasn't entirely what I wanted to do. So then I joined the UNDP in Islamabad. About two years in I was getting bored in the job, as I didn't feel that I was contributing substantively to development. I was mostly doing project management: disbursing funds from the head office, getting the quarterly reports back, checking them and then sending them back to the head office for approval. At the same time I was also organising quite a number of conferences, and realised that I wanted to be doing what the conference attendees were doing, rather than being the conference organiser. They were researching development issues and creating knowledge around which I was simply designing projects. That is when I got interested in pure academia so I went back to the US and then the UK to do a Master's degree, after which I returned to teach in Pakistan at the Lahore University of Management Sciences (LUMS), where I integrated development studies into the political sciences and economics programmes. At the same time I got involved in a lot of field-based primary research in rural Pakistan.

I loved teaching at LUMS but it is impossible to advance in academia without a PhD. So I started one at the Institute of Development Studies (IDS) in Brighton, UK, and after completing the PhD I became a Fellow at the institute. IDS runs a number of Master's programmes and a large PhD programme, so I have teaching and supervisory responsibilities with students but I am involved in conducting my own research at the same time. As IDS is also a development think tank, we have more involvement in policy research and development programmes than those working in purer academic departments in universities. We use our research to guide policy and inform governance in countries across the world and in development pro-grammes. I am also involved in teaching a research methods course to researchers in African universities. My main area of research is the political economy of public policy and service delivery, focused mainly around empirical analyses of decentral-isation and informal institutions.

I like the feeling that we can define the direction of development policy and practice. Donors develop or fund new projects based on *our* literature. It's also a very independent way of working, with a lot of freedom, without being tied to the office. But there are a lot of deadlines, which means working evenings and weekends. When you are not teaching you might be working on your book, or publishing a paper (of which a few are expected each year), writing a proposal for funding, a policy report, or planning your next course.

Shandana Mohmand (Pakistan/Portugal) Kenya,
India, Serbia, Macedonia

RURAL DEVELOPMENT

Despite the global trend towards urbanisation, around 75 per cent of the poor live in rural areas. Even in 20 years, 60 per cent of the poor are expected to live outside of cities.[1] Providing a route out of poverty for these rural residents will remain a priority for national governments and the international community for several decades to come. Rural development is about restructuring the rural economy to improve the living standards and economic opportunities for rural communities. Traditionally, rural development focused on resource extraction such as forestry and agriculture. As the nature of rural areas changes, the focus has been on the rural non-farm economy, engaging creative development approaches such as tourism, niche manufacturers and recreation, although intensifying agriculture without damaging natural resources and the environment remains a critical goal for rural development. Education, entrepreneurship, physical infrastructure and social infrastructure all play an important role in developing rural regions. As rural areas vary enormously, a broad range of rural development approaches are used globally.

The work: The rural poor are quite diverse both in the problems they face and the possible solutions to these problems. You will need to understand these dynamics and engage several actors, especially the communities themselves. You might be using one of several approaches to community change such as Sustainable Livelihoods Approach (SLA), Community Capitals Framework (CCF) and Participatory Action Research (PAR) to the local environment. You might be raising awareness among the local population of the importance of conservation and preserving the natural environment, in order to achieve sustainable development, while maintaining a balance with the history and customs and socio-cultural values within the local community. You could be helping people to access microcredit, engage in economic activities, develop projects that protect the environment while improving people's standards of living, or lobbying for land rights to prevent rural to urban migration. Another focus is on strengthening the community structure and organisations, ensuring that decisions are made in an open and transparent way, for collective, rather than individual interests.

The employers: Rural development is very broad and overarching, encompassing many areas, and thus many organisations have a stake in the field and work towards this goal, either directly or indirectly. NGOs working explicitly on rural development issues include Plan International, World Vision, Save the Children, Oxfam. Projects tend to be no more than five years – anything beyond this time frame can generate dependency within the communities.

Breaking in: A Master's in Rural Development is very common, and is often combined with other issues such as Agricultural Economics, although not all have a developing country focus. An International MSc in Rural Development (IMRD) is one of the prestigious Erasmus Mundus programmes under the framework of the European education system, headed by Ghent University of Belgium with several other European universities taking part.

Job titles: Rural Development Expert; Rural Enterprise Development Coordinator; Cooperative Programme Manager; Rural Water Supply and Infrastructure Advisor

Find out more:

Mary Emery *et al.* (2013) *Sustainable Rural Development: An International Perspective*, London: Routledge.

See also: Livelihoods; Natural resource management; Microcredit; Farming and agribusiness

---

I studied Civil Engineering in Guayaquil, Ecuador, which led me on to work in rural engineering. My first experience of rural development was as a regional coordinator in a credit programme to improve housing conditions in the rural areas in the Guayas Province of Ecuador. With the support from Plan International, families were granted credits to improve their houses using local materials and respecting the traditional architecture. The repayment of the loans generated a seed fund donated to communities to grant new loans for implementing community projects such as water pumps, classrooms and so on. Another one of my roles was in Cochabamba, Bolivia. This was a USAID funded project targeting 56 coca-producing communities. It took over 18 months for the authorities to accept the project, but after this time a plan was made with the communities and six bridges, 12 school classrooms and 29 water systems were built.

Most recently I have been working in Honduras, where we introduced nine cropping systems with supplementary irrigation water at ultra low pressure for the communities of the Valle district. We succeeded in diversifying the crops and enabling them to have a summer harvest – which had not happened previously due to a shortage of water. The communities were assisted in forming water boards, ensuring optimal use of water storage ponds, developing business plans and marketing their products. In some cases the yield of crops increased by 60 per cent through the application of new technology and availability of water for irrigation in the dry season.

Decisions need be based on what will improve the quality of life of its inhabitants at the same time as ensuring environmental conservation. Some of the most common challenges with this type of work include developing development programmes in line with the vision of the communities, as well as bringing about (attitude) change. Sustainability is also a challenge, but strengthening local leaders and bringing about change is a long-term plan with regular follow up.

I enjoy working in the field of rural development because it is targeted to the most vulnerable people. But some of the biggest challenges are ensuring the communities are committed to quality and sustainability. Community leaders and regulations need to adequately ensure social, political and economic develop-ment of the population. It is extremely important to be transparent at all levels of implementation of the work.

Rural populations generally have little education, so communication needs to be at their level. Perseverance is important to break the silence of the population, and avoid generating mistrust. Building confidence within the population requires harmonious, participative and experiential work.

Mario Noboa (Ecuador) Honduras, Bolivia, Peru

Note

1 IFAD, www.ifad.org/rpr2011/report/e/rpr2011.pdf (accessed 30 January 2015).

SHELTER AND HOUSING

The Universal Declaration of Human Rights asserts that adequate shelter and housing are fundamental to our human rights. Yet around the world, thousands of people do not have access to adequate shelter, due to poverty, urban migration, conflict and displacement or natural disasters. In post-disaster humanitarian responses, emergency shelter and settlement assistance is the immediate need to protect people with dignity. After the emergency phase, reconstruction efforts begin to tackle housing, settlement planning, building community infrastructures with water, sanitation and access roads. Areas that are slowly gaining importance, but not at scale, are sustainable futures through energy efficient and green design, accessible to the poorest.

The work: Using limited resources in chaotic times immediately after disaster to shelter the affected population. In this context, you will help deliver shelter/housing projects to some of the most vulnerable communities on earth. The job involves listening to those in need, working out how best one can support them and advocating for their rights, working in some of the harshest environments, with very resilient people. As an architect working in these environments and with local communities, you will assume a very different role to the traditional architectural designer's role. Thus, understanding the political, social and environmental context where you are working is also very important. The scale at which you will work is not comparable to private architectural practice; for example, after the Haiti earthquake (2010) 1.5 million people needed emergency shelter and subsequently permanent homes.

The employers: Some specialist architects practices and consultants focus on humanitarian architecture. International organisations (such as Red Cross, CARE, CRS, OXFAM, Habitat for Humanity (HFH), NRC), and multilateral agencies (such as UNICEF, UN-Habitat, IOM) work in shelter, and will employ a handful of architects to lead their design and reconstructions teams.

Breaking in: You do not need to be an architect to work on shelter projects, but it will help with the technical side. Here, a postgraduate degree in Architecture with professional experience in addition to substantial field experience (gained through volunteer projects or pro bono work) is a huge asset. Architecture sans Frontières (www.asf-uk.org) run education programmes geared towards built environment professionals. Article25 (http://www.article-25.org) based in the UK run an Essentials of Humanitarian Architecture Series. Oxford Brookes University (UK) also runs a Master's in Shelter after Disasters. London Metropolitan University runs a BA and Master's in Architecture of Scarce Resources and Rapid Change. Sheltercentre.org has an internship programme that can give solid understanding of the humanitarian system.

Job titles: Shelter Coordinator/Advisor; Infrastructure Specialist; Community Mobiliser

Find out more:

Emergency Operations and Emergency Response Team (Shelter and Settlements) (2011) *Managing Post-disaster (Re)-Construction Projects: How-to Guide.* Available online at: www.sheltercasestudies.org/files/CRS-managing-post-disaster-reconstruction-projects2012.pdf
S. Hirano (2012) *Learning From the Urban Transitional Shelter Response in Haiti: Lessons From CRS' 2010–2012 Post-earthquake Activities.* Available online at: www.crsprogramquality.org/storage/pubs/.../haiti_shelter_response.pdf
S. Hirano (2012) *Learning From Urban Transitional Settlement Response in the Philippines: Housing, Land and Property Issues.* Available online, www.seep network.org/mafi--the-market-facilitation-initiative--pages-10037.php

See also: Camp management; Engineering

After gaining my Diploma in Architecture at the Architectural Association London, I started working at a private architecture firm where I trained and qualified as a chartered UK architect. In 2003, together with colleagues from the Architectural Association we formed our own private practice (www.if-untitled.com) and also went on to form www.humanitarianschools.org.

I slowly shifted my interest from private sector work to the humanitarian sector and now I have over ten years of experience working for different INGOs, the UN, Cluster, and donors and government in various countries. I moved in this direction because I found that contributing to relieving human suffering after a disaster is hugely rewarding.

I now work as Senior Global Technical Advisor for Shelter and Settlements in the Humanitarian Response Department at Catholic Relief Services (CRS). CRS is a large INGO and is active all over the world. I am the lead for Shelter and Settlement Rapid Response and Recovery (SSRRR). I provide leadership and technical assistance to country programmes on the design, coordination and implementation of shelter and settlement responses. This includes support in assessments, response analysis, design, construction contract administration and monitoring/evaluation/learning. I have been involved in humanitarian shelter and school infrastructure responses since the Asian tsunami and am dedicated to capturing agency-wide learning in SSRRR to take best practices forward.

A typical day: I wake up to BBC news flash; I commute to work in Nairobi where I will check the situation further on ReliefWeb. I check emails to see whether any requests from Country Office have come through for shelter assistance. For example, in March 2012 as the Mali refugee crisis started to unfold, I would be on a conference call to the regional director of West Africa and be on a plane the next day to carry out a rapid assessment at the border of Niger and Mali as refugees poured across. On a mission like this I would wake up at the crack of dawn, hop into a four-by-four Toyota Land Cruiser and travel with an armed escort to the affected area to meet with the community leaders and refugees, to facilitate focus group discussions. Furthermore, I would have a quick walkabout to see what conditions people are living in, agree with the team the most urgent needs and then design a response and head back to the base camp to report back and write up a project proposal in order to get some funds to be released for the project.

My advice to any young architects wanting to work in this field is to start gaining experience working in developing countries early in your career; field experience will help you to understand whether it is your cup of tea or not. The biggest asset you can acquire is the ability to listen to and work with others from totally different backgrounds, values and systems. I repeatedly see the failure of the 'grand vision' traditional archetype in this field – they are just not flexible enough and cannot adapt their role to tackling the real problems on the ground. In parallel to gaining practical experience, get qualified, and know your profession in terms of technical design, contract administration and construction laws. I find that having a solid professional skill set makes it easier for one to contribute at many levels.

Seki Hirano (Japan), Myanmar, Indonesia,
Philippines, Rwanda, DRC

SOCIAL PROTECTION

Social protection encompasses policies and instruments intended to provide individuals, households and communities with adequate mechanisms to manage risks and shocks, such as disasters, illness or unemployment. Adequate risk management and coping strategies in times of distress prevent people from falling (further) into poverty and promote investment in improving livelihoods. Social protection includes safety nets (non-contributory programmes such as cash transfers, in-kind transfers, public works programmes, social pensions and subsidies), social insurance (contributory programmes such as pensions, health insurance), labour market policies (e.g. employment services, training for the unemployed) and social services (e.g. counselling). Social protection systems – the combination of these policies – look different in each country, according to particular development needs. Low-income countries tend to focus more on safety nets to protect the poorest and more vulnerable people. The Oportunidades programme in Mexico and Bolsa Familia in Brazil have lifted thousands of families out of poverty while improving school attendance and health behaviour among their beneficiaries. The international community plays a key role in supporting governments to identify vulnerabilities that pose a threat to development outcomes, identify at-risk populations and help design adequate social protection programmes.

The work: You will be working towards pursuing reforms to enhance the efficiency and sustainability of social protection policies and programmes. This might involve undertaking poverty and vulnerability analysis of communities or segments of the population (e.g. children), identifying constraints to access of basic services and linking with other sectors (health and education), working on labour market policies and employment, and strengthening the capacity of public authorities to design, implement, manage, monitor and evaluate social protection strategies and policies.

The employers: Multilateral organisations such as World Bank and the United Nations have specific departments/units working on social protection. UN agencies such as the ILO, UNICEF and UNWomen work in specific areas of social protection as well as with particularly vulnerable groups. NGOs are quite active in designing and implementing social protection programmes, particularly those related to health insurance or targeting particularly vulnerable groups. Some development consultancies have specialised in the design of certain social protection programmes such as cash transfers.

Breaking in: A background in economics, social policy, development studies or public policy is an advantage. Prior experience in management and organisation of social protection reforms (both social assistance and social insurance) in transition countries is certainly a plus. Experience in relevant sectors such as social services, education, health, food security, disaster risk reduction, skills development and employment policy can facilitate acquiring a comprehensive vision of social protection systems and serve as a good entry point to working in developing countries.

Job titles: Social Protection Advisor; Social Policy Specialist

Find out more:

http://siteresources.worldbank.org/SAFETYNETSANDTRANSFERS/Resources/
    ProtectionandPromotion-Overview.pdf
www.ilo.org/public/english/protection/spfag/index.htm
www.unicef.org/socialpolicy/index_socialprotection.html

See also: Protection; Governance

I studied Economics at university with the ambition of becoming Colombia's Finance Minister. Along the way I changed my dream to becoming Bogota's Mayor so went on to do a Master's in London in Urban Economic Development and Planning. I came back to Colombia to work in a consultancy firm on urban development and social housing programmes, including housing programmes for internally displaced people.

My husband's work moved the family to Cambodia so I started looking for opportunities there. I worked in a consultancy firm in Cambodia but then an opportunity came up to work with the World Bank on Social Protection. My skills in policy analysis and development issues were easily transferable to this area. The initial work involved assisting the government and coordinating development partners in the drafting of a national social protection strategy. This involved identifying vulnerabilities and adequate programmes to tackle them. The drafting and endorsement processes were very successful and now efforts focus on implementing some of the programmes, including a cash transfers programme.

I now live in Luxembourg but remain engaged with social protection in Cambodia as a consultant. I also consult in risk management for the UNDP. I go on missions to Cambodia and support the implementation of the strategy. Having the opportunity of knowing a process from the inside and managing the day-to-day tasks with government and development partners gives you knowledge that only the context can give. As a consultant I now have a fresh perspective that has helped process some of the field knowledge to turn it into more solid technical skills in social protection.

Social protection is a growing area in many countries because there is awareness that economic growth and prosperity are not enough to tackle extreme poverty and inequality. People are not always able to cope with hardships such as unemployment and health shocks and governments understand the need to provide programmes that can protect the vulnerable while promoting adequate behaviours such as savings, insurance and training for the less vulnerable.

What I like most about the work is the way it complements other areas of social policy such as child protection, health, education and social justice. What I like least is the long time it took for social protection to be seen as more than merely 'hand-outs' and be recognised as an important tool for human capital development.

If anyone likes social policy and is interested in taking the side of the beneficiary, rather than the side of the government or service provider, social protection is the area to work in. Social protection is very beneficiary-oriented and it opens the door to those working in social fields to see them from another angle.

Mariana Infante Villarroel (Colombia), Cambodia, Uganda

SUSTAINABLE TOURISM[1]

Tourism is one of the fastest growing industries in the world and ranks among the top three export sectors (and foreign exchange earners) in nearly half of the world's LDCs. It drives economic growth, job creation and development. Sustainable tourism (or ecotourism) is a concept that has grown in popularity in recent years and is now moving into the mainstream. This type of tourism enables the environment and local communities to share the benefit, by prioritising the conservation of nature and community welfare. It is increasingly being seen as a tool for poverty alleviation and development, providing rural communities with a source of income. But it needs to be managed well to ensure that it does contribute to the poorest people's livelihoods, and that they and their environment are not exploited.

The work: From the policy level (the development of a country-wide sustainable tourism masterplan for example), to working on the ground with local communities, developing and implementing projects. When working on a new tourism initiative involving or benefiting local communities, it is important to start by developing a strategic tourism plan. A needs assessment for both the communities and the land must be done to understand the various different outcomes, the political climate of the region, especially in relation to tourism, to identify the key actors, how they work, and what their interest in a tourism project is. You can also find synergies between public and private actors.

The employers: The United Nations World Tourism Organization (UNWTO) is the leading UN agency working on tourism. The World Bank, European Union and UNESCO also fund sustainable tourism programmes. NGOs are also turning to tourism as a source of income generation – both for their projects and the local communities. Many social-enterprise model ecotourism companies also work in this area.

Breaking in: The UNWTO runs a volunteers programme that aims at training young professionals to support initiatives related to the sustainable development of tourism. The theoretical part of this programme involves a university course – Tourism and International Cooperation for Development. A selection of those who complete the university course join the volunteer corps to gain practical experience. Visit www.unwto-themis.org/

Several other courses exist combining tourism and development – for example, Kings College's (London) MA/MSc in Tourism, Environment and Development.

Job titles: Eco-tourism Development Consultant; Ecotourism Marketing Consultant

Find out more:

Ecotourism International www.ecotourisminternational.com
Global Sustainable Tourism Council www.gstcouncil.org
Martha Honey (1998) *Ecotourism and Sustainable Development*, Second Edition: *Who Owns Paradise?* Washington DC: Island Press.
The International Tourism Society www.ecotourism.org
Planeterra Foundation www.planeterra.org
Sustainable Travel International http://sustainabletravel.org

See also: Rural development; Livelihoods

I didn't consciously choose to work in tourism development. It was my interest in tourism and respect for nature coupled with a desire to work closely with communities – for their benefit – that led me to this area. I started off in sports and adventure tourism, leading treks in the Algerian desert for six months a year. I worked for large French tour operators, but realised that there was a demand for sustainable, responsible and solidary tourism.

I started to develop my own circuits, designed for small groups that were sensitive to their local environment, for people who wanted to learn about their environment. As I opened up new routes in the Sahara, I needed guides so I started to train local guides who could lead the circuits.

By doing this I also had to start taking responsibility and be aware of what I was doing. As Aminata Traoré, former Malian Minister of Tourism, said: 'Tourism is like fire, it can boil the pot, but it can also burn the house.' There is nothing more destructive than tourism that is badly planned and managed.

This is how I started to train and build the capacity of future guides and my work expanded into Mauritania, Mali, Madagascar, Niger, Rwanda, Benin. The training involved a combination of practice grounded in theory. At the same time I started to work as an advisor for sustainable and responsible tourism associations which were just becoming fashionable. As a pioneer in the area I have been getting more calls as a consultant to work in the field.

For the last four years I have been working in Benin to create a symbiotic collaboration between ten organisations, both private and NGOs, to work towards the common goal of sustainable tourism.

The Federation of Benin Organization for Responsible Tourism and Solidarity (FBO-TRS) was started in 2008 by people working in ecotourism (NGOs or private sector) who joined forces to strengthen their capacity in this emerging field and have a greater impact. Tourism is often seen as a source of income, but in order for this to happen it is important to create a structure for local people and organisations working in tourism to benefit, and to create networks.

My work has been mainly to highlight the full potential of each structure, and find a common bond. I have been building the capacity of each while creating a strong common structure based on ethics. These are certainly in their infancy, but are innovative forerunners of a new politics of unity and cohesion in International Responsible Tourism. Any policy action is set up to make a common front and reach a goal: the establishment of a permanent system for sustainable tourism development. It's been a huge task, channelling all these ambitions and good intentions and making them work effectively together. It is also challenging to work, and communicate within a culture where the predominant attitude is that 'to share knowledge is to lose power'.

Françoise Widmer (France) Mauritania,
Mali, Madagascar, Niger, Benin

Note

1   Sandal design by Kon Issara from The Noun Project.

URBAN PLANNING

The world is steadily becoming more urban, as people move to cities and towns in search of employment, educational opportunities and higher standards of living. More than half of the world's seven billion population live in towns and cities and by 2030 this number is expected to swell to almost five billion,[1] with urban growth concentrated in Africa and Asia. Mega-cities such as Shanghai, Karachi or Mumbai will rise, although much of the growth is expected to occur in smaller towns that have fewer resources to respond to the magnitude of the change and needs.[2] The dark side is that today one billion people live in urban slums, which are typically overcrowded, polluted, unsafe and dangerous, and lack basic services such as clean water and sanitation. But urban planning is not only critical to provide people with an address in a habitable area with access to basic services to help break the cycle of poverty, it is also critical from a disaster risk reduction and climate change adaptation perspective, including transport and energy efficiency. Urban land-use planning, if led by well-informed policies based on sustainable development principles and supported by well-planned and well-managed initiatives and investments, can help address these challenges.

The work: You may be working on urban growth and urban economy, infrastructure and transportation, land use, social equity and climate change. You will engage others to participate in strategies, studies or surveys, creating public and private partnerships, building capacity or designing plans – from municipal to regional scale.

The employers: The World Bank is taking a lead in this area both in terms of investment and strategy, while the UN is jointly present through UN-Habitat, UNDP (which has an Urban Partnerships for Poverty Reduction Project for example), UNFPA and UNEP. At international level, there are few NGOs and consultancy firms that specialise in urban planning as it is first a governmental obligation. A few worth mentioning are Villes en Transition, GRET and Emergency Architects. There are also many research institutes whose work touches this area.

Breaking in: There are a number of urban planning courses – most people working in the area have a relevant undergraduate or postgraduate qualification. The World Bank E-Institute provides specialised and comprehensive courses on the urban development sector with both World Bank and UN-Habitat experts. Coursera.org can also be a good source and is supported by academic teachers.

Job titles: Urban Planner, Land and Cadastre Specialist, Slum Upgrading Officer, Decentralisation Expert, Housing and Urban Planning Adviser

Find out more:

State of the World's Cities 2010/2011 – Cities for All: Bridging the Urban Divide, www. unhabitat.org/pmss/listItemDetails.aspx?publicationID=2917
State of the World's Cities 2012/2013 – Prosperity of Cities, www.unhabitat.org/pmss/ listItemDetails.aspx?publicationID=3387
World Bank Urban Strategy (2009) *Systems of Cities: Harnessing Urbanization for Growth and Poverty Alleviation*, World Bank Urban and Local Government Strategy, www.wburban strategy.org/urbanstrategy/

See also: Shelter and housing

I trained as an architect assistant, and then architect. After years of working in this field, I felt the need to go back to university to get a specialisation in Urban Planning. I undertook a post Master course on Cities in Development at the French Urban Institute. I later continued my studies through some World Bank e-Institute courses on specific subjects such as Sustainable Urban Land Use Planning, Street Addressing, and Urban Upgrading. I chose to move to the urban scale as I felt I could make a bigger difference and wanted to work internationally to support people in need, discover other countries and learn from different cultures.

I started in Cameroon as an urban expert for the International Volunteer Programme of the French Cooperation. I was based in Yaoundé, in the office of urban development of the City Hall, discovering the gaps and needs in both the administration and the population. This was a real opportunity as junior professional positions are rare in this area of expertise. After this, I had the opportunity to go to Pakistan to support a post-earthquake reconstruction effort with the French NGO Emergency Architects. I was involved in mapping the destruction, undertaking safety assessments, training the Pakistan Army Corps and the architectural design of reconstruction centres for UN-Habitat and a primary school.

I then went on to work with the NGO Villes en Transition in Vietnam as Country Director. Here I was in charge of managing comprehensive and integrated slum upgrading programmes in Ho Chi Minh City and Nha Trang, and developing projects and partnerships in secondary cities in the Mekong Delta region facing climate change and sea level rise. Subsequently I have been involved in short-term consultancies in Cambodia for NGOs and International Development Partners on diverse subjects – urban strategy and policy, informal settlements, social housing, community markets, hospital family houses, urban heritage, floating villages.

Today I am based in Lao PDR, working as Housing and Urban expert with a consulting firm on project proposals for ADB, World Bank, LuxDev and others. This position is very interesting as it enables me to work on a range of projects for different clients from development partners to the private sector. It gives freedom to choose projects and organise my time.

The field of urban planning is very diverse, allowing different areas of focus. It will help if you choose your path early and study accordingly. My background in architecture provides me with a sense of space, but other people working in the area bring an urban economy or urban sociology point of view, for instance. Many of my colleagues started with a background in engineering, geography, economy, politics or sociology before getting an urban specialisation. A Master's degree in Urban Planning can be followed by other short-term specific trainings. Try to plan your career, as recruiters still tend to separate technical from strategic expertise. An area that I think will grow in importance over the coming years is climate change in urban planning; this is getting more and more important so it will have a wide perspective in the near future.

Alain Phe (France) Cameroon, Pakistan, Vietnam,
Cambodia, Laos PDR

Notes

1 www.unfpa.org/pds/urbanization.htm
2 Ibid.

VALUE CHAIN ANALYSIS[1]

The concept of value chains has recently become a development *buzzword* and is seen as a tool to promote economic growth and poverty reduction. It focuses on the range of actors that are involved in the production, processing, distribution, export and retailing process of a product, with each actor seen as interdependent and adding value to the final product. It is especially important in agribusiness. While development organisations now look at the value chain as a whole and partner with the private sector, the primary focus still remains the increase in the income of producers. Fair trade is also linked to the value chain approach as is the M4P (making markets work better for the poor) approach, which sets out the overall framework conditions under which the value chain methodology is applied.

The work: A specialist in this area is able to identify and address sector-specific bottlenecks that hinder an industry, measure potential competitiveness and facilitate market-based interventions. Practically, it might involve training farmers in techniques to improve the quality of their product and address market demands; strengthening relationships between producers, traders and buyers; reducing inefficiencies in the value chain; liaising with banks to offer small enterprises microcredit to improve their products; analysing price movements and economic profitability/competitiveness of the value chain; and talking to governments and mobile operators to develop a market information system through which everyone can access the latest price developments by SMS.

The employers: Most organisations working in agriculture and agribusiness adopt a value chain approach. A small selection of these include ACDI/VOCA (which works to increase cooperatives' productivity in over 100 countries worldwide) and SNV – the Dutch development organisation that works closely with small businesses and cooperatives and emerging sectors within the development sectors to enable market-based support. They also frequently advertise for value chain analysis internships. WFP – Their Purchase for Progress (P4P) – links farmers to the market and by buying food aid locally from smallholders they want to catalyse agricultural development by providing a market; others include ODI (doing research around value chains), IIED, Technoserve (working in cash crops such as coffee).

Breaking in: Any Master's combining development and trade issues will cover value chains, as well as one focusing on agribusiness or agricultural or development economics. In order to specialise in this area, focus on value chains for the dissertation element of your Master's. Alternatively, gain work experience through an internship with an organisation that works on value chains.

Job titles: Value Chain Expert/Advisor

Find out more:

Linda M. Jones (ed.) (2012) *Value Chains in Development: Emerging Theory and Practice*, Rugby: Practical Action Publishing.
Jonathan Mitchell and Christopher Coles (eds) (2011) *Markets and Rural Poverty: Upgrading in Value Chains*. London: Routledge.
SEEP's Market Facilitation Initiative (MAFI), www.seepnetwork.org/mafi--the-market-facilitation-initiative--pages-10037.php
www.globalvaluechains.org/ an initiative from Duke University
www.m4phub.org

See also: Farming and agriculture; Livelihoods; Microfinance

I studied Development Studies at the Centre for International Development (CIDIN) at Nijmegen, Netherlands. After my Master's in Political Sciences I took part in the 'Global Experience Programme', a partnership between the company TNT and the UN World Food Programme (WFP). Through this I undertook an internship with WFP in Zambia and started working on the logistics of distributing food aid to refugee camps and schools. I then got involved with a new programme, Purchase for Progress (P4P), which sought innovative ways to buy everything locally from small-scale farmers instead of shipping food aid from the US. This turned out to be a fascinating experience that introduced me to the world of agriculture and markets. After this I worked as a consultant for the newly established agricultural market commodity exchange in Zambia.

I returned to do a second Master's in International Development, again at CIDIN. The programme combined work experience (four days at a Dutch NGO) and one day a week of lectures. Building on my past experience, after I graduated I joined the P4P programme at the WFP in Rwanda as a Programme Officer.

I now work as a Value Chain Advisor for SNV, a Dutch development organisation, in Ethiopia. I work towards increasing the income of farmers and improving the competitiveness of Ethiopia's sesame (the country's second largest export earner) and grain sector on the world market. I am based in the northern part of Ethiopia, close to where the product is being grown.

While I was never trained specifically in this area, all my working experience has consistently been in value chain development and agricultural markets. In retrospect it may have been better to have specialised in this area at university, but there is a lot you can learn on the job and read about if you are passionate about something.

To give you a peek into a day in my life, in the morning I make my round through the office greeting everyone as social relations come first when working here. I check my emails and discuss the priorities for the day with my team. At ten I have a meeting with a partner organisation. I call some farmer organisations to check how the harvest is coming along and whether they have found a good market for their produce. I also make an appointment with the director of Heineken which just opened up in Ethiopia. I want to discuss whether they can source their raw materials from small-scale farmers in the country. Later that day I drive by the open-air market place and sit on some grain bags to talk to traders about their problems when buying from farmers and selling to companies and consumers in the market. When I get back to the office I continue writing a concept note detailing the problems that beset the sesame and grains value chain and what role SNV can play in brokering better market linkages. I send it to my supervisor and close down my computer as I notice it is already getting dark outside.

Janno van der Laan (Netherlands) Ethiopia, Rwanda, Zambia

Note

1   Chain design by Plinio Fernandes from The Noun Project.

WATER, SANITATION AND HYGIENE (WASH)

WASH, as it is commonly known, (or WatSan) aims to minimise avoidable mortality and morbidity through access to clean drinking water and good sanitation. During a humanitarian response, where infrastructure has been destroyed, or there is a rapid concentration of people in one place, water and sanitation programming, alongside hygiene promotion, is one of the most immediate concerns and key to reducing deaths from water-related diseases. Many refugee camps around the world are not able to provide the recommended minimum daily water requirement of 20 litres per person; while some 30 per cent of camps do not have adequate waste disposal and latrine facilities (UNHCR). But WASH also has a role in development contexts, as the world continues to be off-track to meet the MDG sanitation target. Sub-Saharan Africa and Oceania are considerably below the global coverage of access to clean drinking water (WHO/UNICEF (2014) *Progress on sanitation and drinking-water*. Joint monitoring programme update).

The work: From measuring water quality and planning water treatment to designing a cholera outbreak response or disease surveillance, and overseeing the design features and specifications for boreholes and the work of drilling contractors. As a WASH coordinator you will be overseeing all elements of clean water supply and waste water disposal to reduce disease. Community awareness and training programmes might also be a feature of the work, or you may coordinate with health promotion specialists on this. Here you will promote awareness of the connection between the hygiene practices, poor sanitation, polluted water sources and disease. On a more technical level you may be in charge of installation of surface water treatment systems (SWAT) including community training, borehole rehabilitation, health and hygiene promotion (HHP), and provision of emergency latrines.

The employers: Humanitarian organisations working in the immediate aftermath of a emergency will often focus on WASH. Among these are UNHCR, ICRC, Oxfam. But it is also a priority for development focuses, and WHO, Bill and Melinda Gates Foundation, Centre for Disease Control (CDC) etc. all have programmes in this area.

Breaking in: Engineers, hydrologists and environmental health professionals are well placed to move into WASH programmes with some training. A diploma in either Water Technology (especially drilling techniques) or experience in HHP and sanitation in emergency settings is an advantage. Most of the organisations working in WASH sectors such as UNICEF, IRC, different Red Cross societies and many other NGOs organise different relevant trainings for their staff and organisations in the WASH network but RedR also offers an open course.

Job titles: WASH Manager/Coordinator/Advisor

Find out more:

http://washvacancies.wordpress.com/

See also: Public health

I studied history at Mawlamyine University in Myanmar but became interested in development and humanitarian work after observing the activities of the Red Cross in my native town. I joined the ICRC as a Hygiene Promoter in late 2002. My job mainly involved operation and maintenance of constructed water and sanitation facilities as well as sensitising community awareness on key hygiene messages: hand washing, safe disposal of faeces and drinking safe water. After one year working with grassroots level communities, they really touched my heart. I was welcomed by them and felt appreciated; they shared their food with me although they were poor. I also learned how their lives were vulnerable: struggling in poverty with limited access to markets and education, job opportunities and health care . . . Their villages were remote, living conditions poor: tiny bamboo houses, leaking roofs, no furniture, no latrines and far from water sources. Upon completion of the projects I made a concrete decision to continue the path I had started and specialise in WASH. I like this theme because it offers an interesting mix of public health and engineering and is not limited to management. There is also demand on the job market for WASH professionals. More importantly, it has a direct impact on the lives of the community.

In 2006, after four years of working with the ICRC as a national staff member, I got a scholarship to study at IUED (Institut Universitaire d'Etudes du Développement) in Switzerland. I did a Master's in Development Studies, which allowed me to understand the broad aspects of international development and the crucial role of

WASH. Most of my public health engineering skills were acquired from vocational trainings, practical field experience and self-study.

After I completed my Master's degree I became an international staff member of the ICRC. I also worked for the IFRC and French Red Cross as WASH professional but in different contexts and roles, which means I never tire of the WASH sector. I tried to put myself in different roles such as WASH project manager in conflict and post-conflict contexts, coordinator in emergency response and recovery of large-scale natural disasters, and became involved in long-term WASH development projects but as an advisor. I am currently working for the Médecins Sans Frontières (MSF Holland) in the role of WatSan Advisor based in the HQ. A typical day might include providing technical advice to those countries under my portfolio, developing policy on a specific subject e.g. hygiene promotion), facilitating training and attending meetings.

I like the opportunities this job offers to interact with different people and a wide spectrum of cultures, which help me see things from different angles. There is a lot of job satisfaction due to the big welcome from the beneficiaries and the appreciation of the people around you. Once you work for aid organisations, you can contribute or lead to influencing positive policy changes of the countries. We are also able to see the tangible and immediate results of our efforts in some cases. The cons are that the work is often in insecure countries, and more and more we have become the soft target of arms carriers; it is physically demanding, and compromises our personal life (far from family, no fun, tight security restrictions). I believe it's worthwhile making sacrifices for such people who are in trouble; we stand on their side to make a better world, although it won't be perfect.

My advice to anyone wanting to specialise in this field is to acquire certain technical skills in addition to general public health or engineering, such as water supply – drilling, sanitation – construction of simple pit latrines, hygiene/health promotion training. Entry is not easy, so hunt for jobs in a timely fashion – i.e. make more effort if there is huge emergency because there is often a HR shortage. And be ready for adventure!

Thurein (Myanmar) Uganda, Pakistan, Guyana,
Central African Republic, Chad

APPENDIX 1

Reading lists

Books

What is development?

Starting with a couple of classics, and moving on to some more contemporary writings, this is a short selection of what is out there.

Frantz Fanon (1964) *Towards the African Revolution*, New York: Grove Press.

The Martinique-born French philosopher is very influential in the field of post-colonial studies. Prior to this he also published *Black Skin, White Masks* (1952), *A Dying Colonialism* (1959) and *The Wretched of the Earth* (1961).

Walter Rodney (1972) *How Europe Underdeveloped Africa*, London and Dar es Salaam: Bogle-L'Ouverture.

The prominent Guyanese scholar lays the foundations for the current problem and the Northern Agenda from a southern voice.

Clifford Geertz (1977) *The Interpretation of Cultures*, New York: Basic Books.

This book develops an important new concept of culture and helps define a generation of anthropologists and what their field is ultimately about.

Amartya Sen (1999) *Development as Freedom*, Oxford: Oxford University Press.

Amartya Sen, an Indian economist and a Nobel laureate has worked as Professor of Economics at Jadavpur University Calcutta, the Delhi School of Economics and the London School of Economics, and as Drummond Professor of Political Economy at Oxford University and Harvard University. He has written extensively, and other significant works include *Poverty and Famines: An Essay on Entitlement and Deprivation* (1983), and has also collaborated with other authors.

Martha C. Nussbaum (2001) *Women and Human Development: The Capabilities Approach*, Cambridge: Cambridge University Press.

Prior to this publication, Nussbaum, an American philosopher, collaborated with Amartya Sen in the 1980s and started to address issues of development and ethics culminating in *Quality of Life* (1993), published by Oxford University Press. Her latest book, on capabilities, is *Creating Capabilities, The Human Development Approach* (2011).

Joseph Stiglitz (2010) *Globalisation and its Discontents*, London: Penguin.

Former senior vice president and chief economist of the World Bank, turned critic, Stiglitz is one of the most influential economists and author of several books. This book criticises the management of globalisation and focuses on income distribution and international trade.

On poverty and its elimination

Jeffery Sachs (2006) *The End of Poverty: Economic Possibilities for Our Time*, New York: Penguin.

Labelled as one of the most important economists of our times, Sachs has a plan to eliminate extreme poverty around the world by 2025 and the nine steps to do so are laid out in this book.

Paul Collier (2007) *The Bottom Billion: Why the Poorest Countries Are Failing and What Can Be Done About It*, New York: Oxford University Press.

In this book he offers real hope for solving one of the great humanitarian crises facing the world today. Standard solutions do not work, he writes; aid is often ineffective, and globalisation can actually make matters worse, driving development to more stable nations. What the bottom billion need is a bold new plan supported by the Group of Eight industrialised nations including the adoption of preferential trade policies, new laws against corruption, new international charters, and even carefully calibrated military interventions.

Coimbatore K. Prahalad (2007) *The Fortune at the Bottom of the Pyramid: Eradicating Poverty Through Profits*, Upper Saddle River, NJ: Pearson Education/Prentice Hall.

Collectively, the world's 5 billion poor have vast untapped buying power but are frequently overlooked by corporations. By learning to serve this market and providing the poor with what they need there is enormous potential for companies – and a possibility to reduce poverty.

Nanak Kakwani and Jacques Silber (2008) *The Many Dimensions of Poverty*, Basingstoke: Palgrave Macmillan.

This book takes a multidisciplinary approach to poverty, including five different perspectives from the disciplines of economics, sociology, anthropology, psychology and institutional economics.

Paul Polak (2009) *Out of Poverty: What Works When Traditional Approaches Fail*, San Francisco, CA: Berrett-Koehler Publishers.

This book describes practical solutions to poverty based on the author's over 25 years' experience.

Jacqueline Novogratz (2010) *The Blue Sweater: Bridging the Gap Between Rich and Poor in an Interconnected World*, Emmaus, PA: Rodale Books.

The book is a first-hand account of her journey from international banker to social entrepreneur and founder of Acumen Fund. The book's name comes from a blue sweater, her prized possession, which she gave away to Goodwill; 11 years later she spotted a young boy wearing that very sweater, with her name still on the tag inside in Rwanda. This showed how we are all connected, how our actions – and inaction – touch people every day across the globe. She then went on to set up Acumen, a non-profit venture fund.

Abhijit V. Banerjee and Esther Duflo (2011) *Poor Economics: A Radical Rethinking in the Way to fight Global Poverty*, New York: PublicAffairs.

This important book is by two of co-founders of the Abdul Latif Jameel Poverty Action Lab (J-PAL) – which advocates for using randomized evaluations to study poverty alleviation. In this book they attempt to piece together a coherent story of how poor people live their lives, of the constraints that keep them poor, and of the policies that can alleviate this poverty. Building on the latest evidence drawing on including both experimental and non-experimental evaluations. So they offer a new take on poverty and answer questions like: Why would a man in Morocco who doesn't have enough to eat buy a television?

Dean J. Karlan and Jacob Appel (2012) *More than Good Intentions: Improving the Ways the World's Poor Borrow, Save, Farm, Learn, and Stay Healthy*, New York: Penguin.

Through presenting innovative and successful development interventions around the globe, the authors show how the latest thinking in behavioural economics can make a profound difference.

Paul Polak and Mal Warwick (2013) *The Business Solution to Poverty: Designing Products and Services for Three Billion New Customers*, San Francisco, CA: Berrett-Koehler.

By serving the poor ethically and effectively the authors argue that businesses can earn handsome profits while helping to solve one of the world's most intractable problems.

Aid and its discontents

William Easterly (2002). *The Elusive Quest for Growth: Economists' Adventures and Misadventures in the Tropics*, Boston, MA: MIT Press.

This is Easterly's first book where he criticised the utter ineffectiveness of Western organisations to mitigate global poverty, written while he was employed as a research economist at the World Bank. He was promptly fired. He later also wrote *White Man's Burden: Why the West's Efforts to Aid the Rest Have Done So Much Ill and So Little Good* (2007), where, transformed into an NYU economics professor, he brazenly contends that the West has failed, and continues to fail, to enact its ill-formed, utopian aid plans because, like the colonialists of old, it assumes it knows what is best for everyone. He also has a number of other books worth reading.

Dambisa Moyo (2010) *Dead Aid: Why Aid Is Not Working and How There Is Another Way for Africa*, London: Penguin.

This book reveals why millions are actually poorer because of aid, unable to escape corruption and reduced, in the West's eyes, to a childlike state of beggary.

Samia Altaf (2011) *Too Much Aid, Too Little Development*, Baltimore, MD: John Hopkins University Press.

With case studies from Pakistan this book gives an account of how advisors from overseas go about implementing their 'projects'. It documents the well-documented problems with foreign assistance from within the system.

Emergency contexts

Barbara Harrell-Bond (1986) *Imposing Aid: Emergency Assistance to Refugees*, Oxford: Oxford University Press.

This provides the first independent in-depth appraisal of an assistance programme mounted in response to an emergency influx of refugees following the 1982 crisis involving Ugandans who spilled over the Sudan border.

Rebecca Solnit (2010) *A Paradise Built in Hell: The Extraordinary Communities that Arise in Disaster*. New York: Penguin.

The most startling thing about disasters is not merely that so many people rise to the occasion, but that they do so with joy. That joy reveals an ordinarily unmet yearning for community, purposefulness and meaningful work that disaster often provides.

Conor Foley (2010) *The Thin Blue Line: How Humanitarianism Went to War*, London: Verso.

This describes how in the last 20 years humanitarianism has emerged as a multibillion-dollar industry that has played a leading role in defining humanitarian crises, and shaping the foreign policy of Western governments and the United Nations.

Michael Barnett (2011) *Empire of Humanity: A History of Humanitarianism*, New York: Cornell University Press.

Provides a history of the evolution of humanitarianism.

Inside the everyday lives of development and humanitarian workers

Jill Tweedy (1987) *Internal Affairs*, London: Penguin.

This is a light-hearted fiction book in which she captures the irony of everyday development life in 'Sulenesia'.

Anne-Meike Fechter and Heather Hindman (2010) *Inside the Everyday Lives of Development Workers: The Challenges and Futures of Aidland*, Sterling, VA: Kumarian Press.

This is an edited collection of contributions focusing on the people who *do* development.

James Maskaylk (2010) *Six Months in Sudan: A Young Doctor in a Wartorn Village*, Edinburgh: Cannongate Books.

A field level memoir by expat aid worker.

Ruth Stark and Bill Bicknell (2011) *How to Work in Someone Else's Country*, Washington, DC: University of Washington Press.

An easy-to-read guide offering practical guidance to professional development workers on how to be effective when working with governments and international counterparts.

Simon Worrall (2013) *Under the Paper Mountain: A Journey into the Heart of the United Nations* (Kindle edition only).

A short report about when the author goes undercover at the UN headquarters in New York following the trail of a single document through the intestines of the UN's bloated bureaucracy.

Magazines and journals

Developments – produced by DFID who offer a free subscription to their magazine which aims to raise awareness of development issues. www.developments.org.uk

New Internationalist – exists to report on issues of poverty and inequality. www.newint.org

The Verge: Travel with purpose – Comprehensive resource for those wanting to volunteer, work, study, live or adventure overseas. Canadian magazine www.vergemagazine.com

Pambazuka News www.pambazuka.org

World Development Journal – published by Elsevier. *World Development* is a multi-disciplinary monthly journal of development studies. It seeks to explore ways of improving standards of living, and the human condition generally, by examining potential solutions to problems such as: poverty, unemployment, malnutrition, disease, lack of shelter, environmental degradation, inadequate scientific and technological resources, trade and payments imbalances, international debt, gender and ethnic discrimination, militarism and civil conflict, and lack of popular participation in economic and political life.

Journal of Development Studies – published by Taylor & Francis.

Development Policy Review – published by Wiley.

Development in Practice – published by Taylor & Francis.

Blogs

There are a large number of blogs out there focusing on development by those working in the field and it's one of the most efficient ways to stay up to date with your field. The following list of top ten blogs was compiled by David Girling, lecturer and Director of Research Communication in the School of International Development, University of East Anglia. The list first appeared on his own blog http://social-media-for-development.org/

My Heart's in Accra. Ethan Zuckerman is an academic, blogger and internet activist. He is a senior researcher at the Berkman Centre for Internet and Society and on the board of directors for Global Voices: a global community of citizen media authors. His TED lectures are also superb. www.ethanzuckerman.com/blog/

Social Media for Good. This blog has a lot of practical advice on how to improve communications in international development using digital and social media. http://sm4good.com/

Global Voices is an online global community of bloggers who report on citizen media from around the world. It was founded in 2005 by Ethan Zuckerman and Rebecca MacKinnon, has over 500 contributors and is translated into more than 30 different languages. Its goal is to give voices to those not usually heard in the international mainstream media. http://globalvoicesonline.org/

Social Media for International Aid and Development. A great blog on social media and international development with lots of practical advice. http://some4d.ning.com/

DEV Blog. This is a multi-author blog from the School of International Development, University of East Anglia. The school is a leading global centre of excellence in research and teaching in international development. www.uea.ac.uk/international-development/dev-blog

DFID Bloggers is a multi-author blog from the UK Department for International Development. There are a mixture of group blogs and individual blogs, many from the field. https://dfid.blog.gov.uk/

Poverty Matters. The *Guardian*'s Global Development Blogosphere. It pulls together blog posts from several partners including DFID, ODI, Global Voices, From Poverty to Power and Texas in Africa etc. The great thing about this blog is that it has a wide audience and therefore you get lots of comments. These comments can often be more interesting than the actual blog post. www.theguardian.com/global-development/poverty-matters

From Poverty to Power is written by Duncan Green who is a strategic advisor at Oxfam GB. He is also the author of the book *Poverty to Power* which is where this blog started. Duncan uses his blog to discuss and debate issues from the book. http://oxfamblogs.org/fp2p/

Chris Blattman is an assistant professor of Political Science and Public Affairs at Columbia University. His research examines the causes and consequences of poverty and violence. He is an avid blogger writing about many aspects of international development. http://chrisblattman.com/

Blood and Milk. Alannah Shaikh has worked in international development for over ten years. Alanna believes that international development should be 'efficient, effective and evidence based'. Her posts are eclectic ranging from careers advice to marketing. The blog offers lots of practical advice. http://bloodandmilk.org/

APPENDIX 2

Continuous professional development

Professional development isn't a one-off thing; continuous learning should be part of your career. Make it a priority to assess and improve your skill set, increasing your effectiveness at work, your value to the organisation and enhancing your future career prospects. Continued learning can be self-directed, online through e-courses or short on-site courses. There are also short workshops or online lectures and webinars that can help you gain specific knowledge. There is a huge range of courses available and this is just a snapshot of some of the training institutions and topics available.

Organisations providing broad training

The United Nations University (UNU) is a global think tank and postgraduate teaching organisation headquartered in Japan. While they do run Master's and doctoral programmes they also run a series of shorter non-degree programmes and affiliated programmes in specialist areas. See http://unu.edu/

For UN staff there is also the UN System Staff College which provides a series of short courses in a broad range of areas such as Development, Gender and Human Rights, Development Leadership, Effective Evaluation and Writing Skills. Based in Turin, Italy, some are online courses and others are on-site courses. See www.unssc.org/

INTRAC, the International Humanitarian Training and Research Centre is based in Oxford, UK and provides a range of trainings from Introduction to Evaluation, Advocacy and Policy Influencing, Participatory Proposal Development, etc. Costs range from £595 for a short non-residential course to over £1,295 for a residential course. www.intrac.org/

BOND, based in London runs a number of bespoke and in-house training courses for NGOs in areas such as advocacy, fundraising, programmes and effectiveness. www.bond.org.uk/

Learning in NGOs (LINGOs) has a learning management system (LMS) that contains hundreds of courses, both from their partners and shared by their member agencies, on Leadership and Management Development, Information Technology, Project Management, Stress Management for Humanitarian Workers, Personal Safety and other topics. By providing a community for sharing learning resources and experiences, and the latest learning technologies and courses from our partners, LINGOs helps INGOs increase the skill levels of their employees and thereby increase the impact of their programs. www.lingos.org

Specific training in . . .

Project management

In an industry that relies upon projects to complete its work, specific project management training and/or a certification helps ensure that project managers are ready to effectively manage their projects. There are a number of certification programmes on offer, a selection of which are listed below:

- *PM4DEV* is a Registered Education Provider, approved by the PMI. They provide training services customized to the challenges of development projects with a methodology based on a project management cycle to effectively manage projects of all sizes. They offer the Certified Development Project Manager (CDPM) course as well as individual modules. Their ten-week e-course, Effective Project Management, for example, costs $175 and is worth 15 Professional Developments Units. www.pm4dev.com/
- *PM4NGOs* publishes the free guide *Project Management for Development Professionals* (PMD Pro) and also offers a three-level certification scheme for Project Managers working in the development sector. The scheme is aligned with existing internationally recognised project management standards and includes components specific to the NGO sector.www.pm4ngos.org/
- The Cooperation & Development Centre (CODEV) of Lausanne offers the Certificate of Advanced Studies in Management of Development Projects (MaDePro) targeted at project managers, engineers, architects, social scientists and other equivalent university graduates. It is a 33-week course with an e-learning component, a two-week field trip, and an individual project, costing just over $4,000. Grants are available for participants from developing or emerging countries. http://cooperation.epfl.ch/madepro
- Monterey Institute of International Studies offers Certificate Training in Development Project Management. Their three-week intensive DPMI programme focuses on creating programmes, facilitating stakeholder engagement, and strategic partnering to meet current development challenges, and is hosted every January in Monterey, California and Kigali, Rwanda and every May/ June in Monterey and Washington, DC. Cost is around $2,500. www.miis. edu/academics/short/development-management

Monitoring and evaluation

Most development initiatives take place in highly dynamic contexts. The change process is complex and uncertain. Guiding a project towards impact will require ongoing monitoring and evaluation and results-based management. Generalists also need to have specific knowledge in M&E.

e-Learning on Development Evaluation. UNICEF and the International Organisation for Cooperation in Evaluation (IOCE), under the EvalPartners initiative, offer an introductory e-Learning programme on Development Evaluation. They also run free webinars where you can watch live presentations and have the option to ask questions and provide comments. These events enable the sharing of good practices and lessons learned on designing and implementing national and local M&E systems. One such webinar is emerging practices in development evaluation. The website also lists a whole range of other courses linking to monitoring and evaluation from self-study to PhD courses and provides access to M&E handbooks, manuals and a roster for M&E professionals. http://mymande.org/elearning

TrainEval. This is a course in development evaluation with special attention to the evaluation approach of the European Commission (EC). TrainEval imparts evaluation theory, applied knowledge and the abilities for planning, conducting and steering development evaluations. It contains four modules of four days each and costs €3,980 (four modules). It is a physical course conducted in Brussels, Belgium. http://traineval.org/

Many universities run short courses in monitoring and evaluations. Wageningen University provides one such course. www.wageningenur.nl/en/show/CDIcourse _PPME_2014.htm and you might be eligible for a Nuffic short courses scholarship www.studyinholland.nl/scholarships/scholarships-administered-by-nuffic/nether lands-fellowship-programmes

Humanitarian action

RedR UK runs an introductory course 'So You Think You Want to Be a Relief Worker?' It is a one-day workshop that provides an introduction for anyone interested in a career in the humanitarian sector. In addition they run a whole series of other short training courses targeting the humanitarian sector such as 'Introduction to Water Supply in Humanitarian Emergencies and Disaster Management'. http://redr.org.uk/en/Training-and-more

The Center for International Humanitarian Cooperation (CIHC) was founded in 1992 to promote healing and peace in countries shattered by natural disasters, armed conflicts, and ethnic violence. They promote innovative educational programmes and training models and have an extensive list of publications and regular symposia that address both the basic issues and the emerging challenges of humanitarian assistance. They run an international Diploma Course in Humanitarian Action, and other courses such as Disaster Management Training and Mental Health in Complex Emergencies. www.cihc.org/

ALNAP strengthens Humanitarian Action through Evaluation and Learning. Their website includes a number of publications primarily addressing the effectiveness of humanitarian interventions. www.alnap.org/

Enhanced Learning and Research for Humanitarian Assistance (ELRHA) aims to stimulate and support collaborative partnerships between academic and humanitarian communities to produce research and training that delivers measurable impact in the prevention of and response to global humanitarian crises. Through their website you can link to a range of courses and centres. www.elrha.org/courses-and-centres

Fundraising and grant management

As there is greater competition for grants and tenders on one side, and on the other donors lay out more stringent guidelines to managing their grants, staff need to be up to date with the latest developments. Developing knowledge in these areas and keeping up to date is important.

Funds for NGOs runs a series of webinars such as How to Raise Funds from the European Commission and specific training videos such as How to Raise Funds for Human Rights Issues. www.fundsforngos.org/

The Institute for Grant Management based in Kenya runs a training course in the grant management cycles: www.infogrant.org/

Inside NGO runs regular workshops on various aspects of managing USAID funding such as: 'An Introduction to USAID Proposal Development'; 'USAID Rules & Regulations: Grants & Cooperative Agreements'; 'Financial Management for US Government Funding' and many others. Training is run at various locations worldwide. www.insidengo.org

NGO Connect produces 'The Essential NGO Guide to Managing Your USAID Award' and a 'Guide for USAID Implementation Partners'. www.ngoconnect.net/resources/ngoguide

This section has just provided a brief snapshot of some of the trainings available out there but there is much more available. Once you have a clear idea of areas that you want to professionally develop in, specifically research the options and training available.

APPENDIX 3

Humanitarian competencies

See Figure A3.1 on the next two pages.

Core Humanitarian Competencies Framework

Keeping crisis-affected people at the centre of what we do

cbha
The Consortium of British Humanitarian Agencies

Competency Domains	Understanding humanitarian contexts and applying humanitarian principles	Achieving results	Developing and maintaining collaborative relationships	Operating safely and securely at all times	Managing yourself in a pressured and changing environment	Demonstrating leadership in humanitarian response
Outcomes	Understand operating contexts, key stakeholders and practices affecting current and future humanitarian interventions	Be accountable for your work and use resources effectively to achieve lasting results	Develop and maintain collaborative and coordinated relationships with stakeholders and staff	Operate safely and securely in a pressured environment	Adapt to pressure and change to operate effectively within humanitarian contexts	Demonstrate humanitarian values and principles, and motivate others to achieve results in complex situations, independent of one's role, function or seniority
Competencies and Core Behaviours for all staff in humanitarian response, informed by skills and knowledge	● **Understanding the humanitarian context** › Demonstrate understanding of the phases of humanitarian response including preparedness and contingency, Disaster Risk Reduction, response and recovery. › Apply understanding of the political and cultural context and underlying causes of the humanitarian crisis. › Demonstrate understanding of the gender and diversity dimensions of humanitarian situations. › Take into account the needs, skills, capacities and experience of crisis-affected people and apply these in the response. ● **Applying humanitarian standards and principles** › Ensure that programme goals, activities and staff	● **Ensuring programme quality and impact** › Demonstrate understanding of agency project cycle management. › Actively participate in the design and implementation of effective projects and programmes. › Maintain focus on delivery of timely and appropriate results using available resources. ● **Working accountably** › Be answerable to crisis-affected people for your actions and decisions. › Collect, analyse and disseminate relevant and useful information and feedback with crisis-affected people and other stakeholders. ● **Making decisions** › Demonstrate flexibility to	● **Listening and creating dialogue** › Actively listen to new and different perspectives and experiences of crisis-affected people, stakeholders and team members. › Establish and maintain clear dialogue with crisis-affected people or other stakeholders. ● **Working with others** › Contribute positively in the team to achieve programme objectives. › Share useful information and knowledge with colleagues, partners and crisis-affected people as and when appropriate. › Actively participate in networks to access and contribute to good practice. › Challenge decisions and behaviour which breach the	● **Minimising risk to communities, partners and stakeholders** › Pay attention to the safety of crisis-affected people and other key stakeholders Identify and communicate risk and threats and mitigate these for you and your agency. › Take measures to 'do no harm' and to minimise risks for your partners and the crisis-affected people you work with. ● **Managing personal safety and security** › Build and sustain acceptance for your work in line with humanitarian principles and standards. › Reduce vulnerability by complying with safety and security protocols set by your organisation and adapt them to the local context.	● **Adapting and coping** › Remain focused on your objectives and goals in a rapidly changing environment. › Adapt calmly to changing situations and constraints. › Recognise personal stress and take steps to reduce it. › Remain constructive and positive under stress to be able to tolerate difficult and challenging environments. ● **Maintaining professionalism** › Take responsibility for your own work and its impact on others. › Plan, prioritise and perform tasks well under pressure. › Maintain ethical and professional behaviour in accordance with relevant codes of conduct.	● **Self-awareness** › Show awareness of your own strengths and limitations and their impact on others. › Demonstrate understanding of your skills and how they complement those of others to build team effectiveness. › Seek and reflect on feedback to improve your performance. ● **Motivating and influencing others** › Communicate humanitarian values and encourage others to share them. › Inspire confidence in others. › Speak out clearly for organisational beliefs and values. › Demonstrate active listening to encourage team

FIGURE A3.1 Core humanitarian competencies framework.

collaboration.
> Influence others positively to achieve programme goals.

● Critical judgement
> Analyse and exercise judgment in challenging situations in the absence of specific guidance.
> Demonstrate initiative and suggest creative improvements and better ways of working.
> Demonstrate tenacity to achieve results.

> Demonstrate personal integrity by using one's position responsibly and fairly.
> Be aware of internal and external influences that affect your performance.

> Champion the importance of safety and keep the safety of colleagues and team members in mind at all times.

International Red Cross and Red Crescent and NGOs / individual agency Codes of Conduct.

adapt in situations of rapid change, always informed by a focus on crisis-affected people.
> Demonstrate understanding of when a decision can be taken and when to involve others.
> Consider the wider impact of your decisions in order to achieve results.

behaviour uphold key national and international humanitarian frameworks, standards, principles and codes which your organisation has committed to.
> Use your power responsibly, in line with accountability principles and standards.
> Demonstrate understanding of your role and that of your organisation and others within the humanitarian system.
> Demonstrate an understanding of coordination mechanisms.

Additional Behaviours for 1st level line managers in humanitarian response, informed by skills and knowledge

Responsibilities typically include:
• leading a functional team
• managing operational delivery
• line management
• budget and resource management

● Understanding the humanitarian context
> Assess and analyse key issues in the humanitarian situation and formulate actions to respond to them.

● Applying humanitarian standards and principles
> Participate in the development of an organisational response based on an understanding of the operating context.
> Respect International humanitarian law and relevant treaties.
> Actively participate in disaster coordination and interagency cooperation, based on a clear understanding of your organisation's perspective and approach.

● Ensuring programme quality and impact
> Set standards in your work and follow agreed operating procedures.
> Clarify roles and responsibilities within your team to maximise impact.
> Collaborate with stakeholders to avoid duplication and maximise resources.
> Regularly provide feedback and information to achieve improved results.
> Document lessons learned and apply them to future projects.

● Working accountably
> Establish processes through which crisis-affected people can participate in the response and share their expectations and concerns
> Ensure efficient and transparent use of resources in accordance with internal controls.

● Listening and creating dialogue
> Ensure feedback from crisis-affected people, partners and other stakeholders is incorporated into programme design, implementation and learning.

● Working with others
> Establish clear objectives with teams and individuals
> Monitor work progress and individual performance.
> Establish agreed ways of working at a distance with partners and staff.
> Work with your team to build trust with communities and stakeholders.
> Foster collaborative, transparent and accountable relationships through partners to formalise and implement partnering agreements.
> Use negotiation and conflict resolution skills to support positive outcomes.

● Minimising risk to communities, partners and stakeholders
> Undertake effective risk assessments with crisis-affected people and partners.
> Demonstrate an understanding of wider UN/NGO security co-ordination and how your organisation participates in those mechanisms.
> Develop contingency plans.

● Managing personal safety and security
> Monitor security risks and ensure organisational protocols are understood and consistently followed by staff.
> Take appropriate action and provide direction and support to team members in the event of a crisis.

● Adapting and coping
> Help others to recognise and manage their own stress by modeling appropriate self care and prioritising your workload.
> Promote well-being and a 'duty of care' culture.

● Maintaining professionalism
> Set realistic deadlines and goals.
> Enable others to carry out their roles and responsibilities.
> Monitor commitments and actions transparently
> Take time to learn from experience and feedback and apply the learning in new situations.

● Motivating and influencing others
> Inspire others by clearly articulating and demonstrating the values, core purpose and principles that underpin humanitarian work.
> Provide regular and ongoing informal and formal feedback.
> Recognise the contribution of others.
> Adapt leadership style to the time frame and changing situation.

● Critical judgement
> Maintain a broad strategic perspective at the same time as an awareness of the detail of a situation.
> Act decisively and adapt plans quickly to respond to emerging situations and changing environments.
> Take informed and calculated risks to improve performance.

INDEX

Page numbers in *italics* denote an illustration/figure/table